Lecture Notes in Economics and Mathematical Systems

Managing Editors: M. Beckmann and W. Krelle

306

Christoph Klein

A Static Microeconomic Model of Pure Competition

Springer-Verlag
Berlin Heidelberg New York London Paris Tokyo

Author

Dr. Christoph Klein
Institut für Statistik und Mathematische Wirtschaftstheorie
Universität Karlsruhe
Kollegium am Schloß, Bau III, D-7500 Karlsruhe 1, FRG

ISBN 3-540-19358-8 Springer-Verlag Berlin Heidelberg New York
ISBN 0-387-19358-8 Springer-Verlag New York Berlin Heidelberg

Printed in Germany

Printing and binding: Druckhaus Beltz, Hemsbach/Bergstr.
2142/3140-543210

FÜR MEINE ELTERN

ZUR

GOLDENEN HOCHZEIT

PREFACE

The article presented here reflects a scientific work of more than ten years. During this too long period I enjoyed useful discussions with many researchers and scientists. I offer my apologies for being unable to thank every single one of them here explicitly. Valuable hints of Dr.A.Klodt, Prof.Z.Lipecki, and Prof.B.Peleg improved the work. The helpful conversations with and the hospitality of Prof.T.E.Armstrong and Prof.M.K.Richter are gratefully acknowledged. The instructions in macroeconomics by Prof.G.Gabisch were very helpful for my principal reasoning. I am indebted to my teacher and supervisor Prof.D.Pallaschke, who motivated and supported always our research. Especially I have to thank Prof.S.Rolewicz from the Polish Academy of Sciences for the great help and encouragement.

The first three chapters of this paper correspond to my Habilitationsarbeit submitted to the Faculty of Economics of the University of Karlsruhe in February 1987.

I am indebted to Prof.W.Eichhorn, in particular for his seminars. The effective support by our chairman Prof.R.Henn is gratefully acknowledged.

I wish to express my gratitude and my admiration to Prof.Dr.h.c. mult.W.Krelle, editor of this scientific series.

The excellent cooperation with Springer Verlag Heidelberg is gratefully acknowledged. I am indebted to Mrs.K.L.Stephan for improving my english. The support of our institute in preparing the typed manuscript is acknowledged with thanks.

March 1988 C.Klein

CONTENTS

INTRODUCTION

In the paper presented here we elaborate a static, microeconomic model of a market with pure competition. The status of pure competition holds in an exchange market if no single member of the market can influence the market situation similar e.g. to how a single voter can rarely influence the presidential election.

Although every market model is an economic abstraction, a microeconomic model of a market with pure competition is necessarily a strong idealization: If we require in the idealized case that no single member possesses a positive market influence then we have to introduce markets with infinitely many members, called agents, to model the status of pure competition. Indeed exchange markets with pure competition are an instrument of the microeconomic theory comparable to the notion of a pure vacuum or an ideal gas in physics.

Beside monopolistic and oligopolistic markets the status of pure competition is one out of three principal shapes of markets studied within the economic theory. Hence within the microeconomic research there is a natural need of a precisely interpretable and easy-to-handle model of a market with pure competition.

The study of markets with pure competition was often motivated in the past by the wish to verify F.Y.EDGEWORTH's conjecture [10]. This conjecture states that the set of Walras allocations approximates precisely the Core provided that the set of agents is tending to infinity. H.SCARF [32], and G.DEBREU [8] verified the conjecture concerning markets which are countably infinite replications of finite markets. Due to the rather formal construction of these replicated markets, the approach was not completely satisfactory. In 1964 R.J.AUMANN [5] verified the conjecture within a market model which used the unit interval [0,1] as the space of agents. The σ-additive Lebesgue measure was chosen to model the natural influence of coalitions. In Aumann's model it was difficult to describe precisely the meaning of an agent and of a coalition. This was caused by problems arising within the foundations of probability theory studied e.g. by J.LOS [26]. R.J.Aumann interpreted an agent as an infinitesimal small interval similar e.g. to how this notion was used in classical mechanics within the last century. In 1978 D.PALLASCHKE [30] tried to combine both approaches. He used the natural numbers $\mathbb{N}$ as the set of agents and introduced the natural influence measure by a finitely additive measure. Unfortunately the finitely additive measure theory was not perfectly developed at this time.

Before we continue to discuss the wide range of the existing literature concerning markets with pure competition we want to present the major ideas and principles of our model. The reason is that our model, based and motivated by D.Pallaschke's approach, is situated between the Debreu-Scarf model and the model of R.J.Aumann. Indeed our attempt can be seen as a completion of the Debreu-Scarf approach using the Jordan content-Riemann integration theory. This old measure theoretic tool was recently recognized by S.ROLEWICZ and the author [15], [16] to be equivalent to the finitely additive measure theory applied by D.Pallaschke.

The presented paper consists of three chapters. In the first one we observe a market, having a set N of countably infinitely many agents, by methods of measurement which are closely related and motivated by *strictly microeconomic measurement tools*, i.e. total statistical censuses, microcensuses and public opinion polls. The observation approach is a *deterministic* one and its logical implications for modelling the situation of pure competition are studied. The natural comparability to the finitely additive approach is investigated. The resulting observed market is a *continuous counterpart* of the original discret market. It can be seen as a *scanning picture* of the original market.

In the second chapter we combine the observation principles based on microeconomic measurement with the *mathematical completion* concept; i.e. we interpret our model additionally as the completion of all replicated markets. This combination allows us to obtain a concrete, deterministic model of a market with countably infinitely many agents. The model consists mainly of a continuous mapping from a well determined and precisely explained compact space into the space of the agent's characteristics. The countably infinite set N of agents is a dense subset of this compact space. A generalized Jordan content on this compact space yields the natural influence measure which can be explained by a limit of average means.

The nature of pure competition in microeconomics as well as the measurement by public opinion polls require a *probabilistic reinterpretation* of the deduced, deterministic, and continuous model. This reinterpretation is performed by considering equivalence classes concerning the *Jordan content-Riemann integration* theory. Motivated by problems arising in functional analysis [31] S.ROLEWICZ has developed together with us this theory in a generalized manner [15], [16], [17], [18]. The theory of uniformly distributed sequences presented in the monographs of L. KUIPERS and H.NIEDERREITER [24], and of E.HLAWKA [14] allows us to treat the resulting model on the mentioned compact space or equivalently on the set N of agents. At least so far as the nature of pure competition is involved the resulting, *static model of a market with pure competition* seems to us to be explained in every detail. It is based on *microeconomic* methods of measurement.

At the costs of slight losses within the interpretability we could represent the elaborated model by a series of mathematical tricks in an *elementary* manner. The mathematical requirements are reduced on the classical Jordan content-Riemann integration theory and on the definition of continuity. The rational numbers in the unit interval [0,1) are the agents, the rational intervals are the coalitions.

The first two chapters are considered by us to be the most important part of this research paper, namely the presentation of a well interpreted, static, microeconomic model of a market with pure competition.

In the third chapter we probe the aptitude of the developed model with respect to applications within the classical, microeconomic equilibrium theory.

The appendix recalls some technical proofs and tools used within the text. In particular a unified rendering of the elementarily represented model is gathered from the various parts of our article.

We continue to discuss the literature dealing with large economies. The problem is studied in the monographs of R.J.AUMANN and L.S.SHAPLEY [6], W.HILDENBRAND [13], and A.MAS COLELL [27]. We are not able here to list all recent papers. Instead of this we refer the reader to the good survey article of T.E.ARMSTRONG [1]. However it is necessary to report that T.E.ARMSTRONG and M.K.RICHTER [2], [3] recently installed a very general model, based on the finitely additive theory, which includes all previous approaches. Necessarily the model is more abstract and less specialized than the model presented here. There is also the book of M.MORISHIMA [28]. Even if it may be considered to be not fully satisfactory from a rigorous mathematical viewpoint, it seems to us to reflect, better than other contributions, the final microeconomic goal within this topic, namely the installation of a model allowing not only a static but also a comparative static and dynamic microeconomic analysis in a manner well to interpret from the viewpoint of reality.

CHAPTER I. THE DETERMINISTIC GROUNDING OF THE MODEL

In this chapter we will develop the principles, properties, and consequences of an idealized observation of a large market with pure competition. The observation is necessarily an idealized one, but we have carefully elaborated its correlation to practical measurement methods. With the help of our observation procedure we deduce a continuous counterpart of the original large market. The theoretic equivalence of both concepts is deduced. Elements of probability theory are not involved in the first chapter. They will be combined with the model developed here in the second chapter. Our goal is to install an observation approach closely related to practical measurement methods and to scrutinize its principal and logical implications for modeling a large market with pure competition.

Since the first chapter is a principal and general one we could not avoid an abstract mathematical language. The reader less familiar with our framework should not be worried. Firstly, we reduced technical considerations by a systematic use of references. Secondly, the mathematical framework is needed in further chapters only with respect to a concrete and much easier to understand example. Provided that the reader trusts in our exactness and proofs he can understand the logical meaning due to the verbal explanations - that is our intention at least.

The reader interested in technical details can find in the appendix several proofs which may be less familiar and which are refered here only.

1. THE OBSERVATION OF A SET OF AGENTS

A set consisting of a large number of agents is considered. We want to obtain information about the set of agents by public opinion polls or by statistics from a census. This kind of observation is not optimal to scrutinize an individual. It is suitable for studying groups. As we are interested in the case of pure competition, it is in our opinion the appropriate mean. Pure competition holds if the influence of every single agent on the economic situation is zero.

1.1 OBSERVABILITY PRINCIPLES. In preparing a census or a public opinion poll, one acts as follows: At first one defines precisely what is to be scrutinized, i.e. one defines a special question or a special kind of statistical data. Then one fixes the exactness by which the answer has to be registered. Hence the exactness by which the answer is registered is independent on the individual person examined. It is dependent on those persons who plan the public opinion poll or the census (cf. E.NOELLE-NEUMANN [29]).

We assume that only *finite, discrete scales of measurement* are used. We deem this to be a realistic assumption. Real valued scales or unbounded scales are mathematical idealizations in our opinion. When the question and the scale of measurement is fixed, one begins to question resp. begins to collect the statistical data. We formulate this observation process in the abstract mathematical language:

A *method of measurement* is a map $\varphi_G: N \rightarrow \{0,1\}$. N is the *set of agents*, $\{0,1\}$ is the *scale* of measurement. $G_\varphi = \{n \in N \mid \varphi_G(n) = 1\}$ is the measured *group* of agents. From the viewpoint of an agent $n \in N$ this means that he has a *property* such that the value of $\varphi_G(n)$ can be decided. We allow only such groups which can be identified by a method of measurement. Therefore, we treat a *method of measurement*, a *property*, and a *group* as equivalent notions of the same concept.

The above definition is a mathematical simplification and idealization due to several reasons: (i) Generally one uses scales of measurement other than $\{0,1\}$. One can reduce every method of measurement with a finite scale to finitely many methods with the scale $\{0,1\}$. We will admit later on finite combinations of methods of measurement. Therefore we will use exclusively methods possessing a $\{0,1\}$ scale from now on. (ii) If $n \in N$ then we assume that $\varphi_G(n)$ is decided correctly. This is an idealization. Indeed a method of measurement is an idealized process of measurement which is deduced from instruments of measurement used in reality. Since we want to establish an idealized model - i.e. a model for a market with pure competition - we deem this abstraction to be permissible. (iii) Setting aside the simplificated $\{0,1\}$ scale, every public opinion poll is based on a method of measurement (cf. E.NOELLE-NEUMANN [29]). In a public opinion poll one calculates not the group G_φ itself but only the fraction or the relative cardinality (=(cardinality G_φ)/(cardinality N) if N is finite). Therefore a method of measurement is much more related to a total statistical census. Indeed φ_G is a *deterministic* instrument of measurement. We will not use probabilistic elements before chapter 2.

Let $\Phi := \{\varphi_i \mid i \in I\}$ a family of methods of measurement where I is an index set. If it is possible to measure every single group G(i), then we are able to measure *finite combinations of groups* G(i), $i \in I$ by finite combinations of the related methods of measurement φ_i. A finite combination of groups is understood as a finite Boolean polynomial consisting of finitely many complement, intersection, and union operations. A group $A \subset N$, is called *Φ-observable* if it can be obtained by a finite combination of the original groups G(i). In the mathematical formulation we obtain:

If $\Phi := \{\varphi_i \mid i \in I\}$ is a family of methods of measurement then the related groups G(i) generate the *Boolean algebra* BG(Φ) *of Φ-observable* *groups* in N. BG(Φ) is a Boolean subalgebra of the *power set* $\mathbb{P}(N)$ of N which is a complete Boolean algebra. If for n, n'$\in$N, n$\neq$n' there is always an A(n,n')$\in$BG(Φ) with n$\in$A(n,n'), n'$\notin$A(n,n') then Φ *separates* N.

We mention some points: (iv) Infinite combinations of methods of measurement are *not* considered since no individual can answer infinitely many questions. (v) The assumption that finitely many questions can be combined is an idealization. In reality only combinations of few questions are possible due to the restricted time to answer. (vi) The Boolean algebra BG(Φ) is an idealized concept characterizing all groups which *are possible* to observe. We do not imagine that they are all observed simultaneously.

We recapitulate the three main principles of the observation process: (A1) The set of agents is observed in a *deterministic* way (A2) *Finite scales* are used only (A3) Only *finite combinations* of methods of measurement are admitted.

1.2 STONE SPACES AND COMPACTIFICATIONS. Here we describe a first step from a discrete model to a continuous one.

An agent n$\in$N is characterized with respect to the system Φ of methods of measurement by his *property combination* $(\varphi_i(n))_{i\in I}$. This characterization is in general a theoretical concept since we excluded questioning an individual infinitely. We call a zero-one system $(a_i)_{i\in I}$, $a_i\in\{0,1\}$ a *realized* property combination if there exists an n$\in$N with $(a_i)_{i\in I}=(\varphi_i(n))_{i\in I}$. If Φ separates N then the following mapping is injective: $n\mapsto(\varphi_i(n))_{i\in I}\ \forall n\in N$.

The set of *all possible* property combinations with respect to the system Φ is $\{(a_i)_{i\in I}\,|\,a_i\in\{0,1\}\}=\{0,1\}^I=2^I$. With respect to the product topology, the space 2^I is well known as the compact, zero-dimensional Cantor-I-cube. Here a compact space is called zero-dimensional iff it has a neighbourhood basis system consisting of sets which are closed and open (abbreviated: *clopen*). Observe that the finite union, the finite intersection, and complements of clopen sets are clopen.

If $io\in I$ then we call $\bar{G}(io):=\{(a_i)_{i\in I}\mid a_i\in\{0,1\},\ a_{io}=1\}\subset 2^I$ the *property group* with respect to io. The system $\bar{G}(i)$, $i\in I$ generates the Boolean subalgebra $\overline{BG}(\Phi)$ of $\mathbb{P}(2^I)$. The Boolean algebra $\overline{BG}(\Phi)$ coincides with the class $CO(2^I)$ of *all clopen subsets* of the Cantor-I-cube 2^I. Especially $\overline{BG}(\Phi)$ is a neighbourhood basis system of 2^I (cf. Z.SEMADENI [33], § 8 for details).

If $\bar{A}\in\overline{BG}(\Phi)$ then $\bar{A}$ is a Boolean polynomial of suitable property groups $\bar{G}(i)$. Let $A\subset N$ be the same Boolean polynomial performed with the related groups $G(i)$. One verifies that $\bar{A}$ contains no realized property combination iff $A=\emptyset$. Hence, by a *finite* combination of the considered methods of measurement we are able to verify whether $\bar{A}$ contains no realized property combination. Therefore we deem the Boolean ideal $\mathcal{Z}:=\{\bar{A}\subset\overline{BG}(\Phi)\mid A=\emptyset\}$ to be superfluous for a description of N by Φ.

If $\bar{A}\in\overline{BG}(\Phi)$ then $\bar{A}$ is a clopen set. Therefore $Z:=\cup\{\bar{A}\mid\bar{A}\in\mathcal{Z}\}$ is open in 2^I and $2^I\setminus Z$ is a compact, zero-dimensional space. We claim that $2^I\setminus Z$ is the *Stone space of* $BG(\Phi)$ denoted $Stone(BG(\Phi))$. Indeed $BG(\Phi)=\overline{BG}(\Phi)/\mathcal{Z}$ and therefore $Stone(BG(\Phi))=Stone(\overline{BG}(\Phi))\setminus Z=2^I\setminus Z$. The clopen sets $CO(Stone(BG(\Phi)))$ of $Stone(BG(\Phi))$ are $\{\bar{A}\setminus Z\mid\bar{A}\in\overline{BG}(\Phi)\}$ and they form a neighbourhood basis system of $Stone(BG(\Phi))$ (cf. R.SIKORSKI [35], § 8 and [21] for details and proofs). Due to our interpretation *every point and every clopen subset of* $Stone(BG(\Phi))$ *has a well determined meaning.* Indeed, every point of $Stone(BG(\Phi))$ corresponds to a possible combination of methods of measurements. Every clopen subset corresponds to a measurement by a finite combination of methods of measurement. Points in Z resp. elements of $\mathcal{Z}$ are neglected.

If $A\in BG(\Phi)$ then one easily verifies that the realized property combinations $(\varphi_i(n))_{i\in I}$, $n\in A$ are densly contained in the related clopen subset $\bar{A}\setminus Z$ of Stone(BG(Φ)). We can naturally identify N with the set of realized property combinations if we suppose that Φ separates N. Then N is a dense subset of Stone(BG(Φ)) fulfilling $\bar{A}\setminus Z$=cl(A) for every $A\in BG(\Phi)$. Here cl(A) denotes the closure of A. On the other hand, if $B\subset$ Stone(BG(Φ)) is clopen then $B\cap N\in BG(\Phi)$. Therefore Stone(BG(Φ)) can be interpreted as a *compactification* of N. Remember that a compactification of a topological space X is a compact space K such that X can be interpreted as a subspace of K. In our case we interpret N as topological space where BG(Φ) is seen as the basis consisting of all clopen sets e.g. if $N=\mathbb{N}$, $G(i)=\{1,\ldots,i\}\ \forall i\in\mathbb{N}=I$ then Stone(BG(Φ))=$\gamma\mathbb{N}$. Here $\gamma\mathbb{N}$ is the one point or Alexandroff compactification of $\mathbb{N}$ (cf. [21] 1.2 for details and further examples).

1.3 THE EXTENSION OF A FINITELY ADDITIVE MEASURE. If N is a finite set then the fraction or the *natural influence measure* is canonically delivered by the Laplace probability measure: If $A\in BG(\Phi)$ then (cardinality of A)/(cardinality of N) is the natural influence of the group A.

Let N be infinite. In this general part we can not offer a motivation or a conclusion which yields the influence of every observed group in a natural way. Instead of that we *suppose* in chapter one that the natural influence of groups is known and that it is given by a non-negative, normalized, finitely additive measure ν, i.e. we assume that $\nu: BG(\Phi)\rightarrow[0,1]$ fulfills $\nu(\emptyset)=0$, $\nu(N)=1$ and $\nu(A\cup B)=\nu(A)+\nu(B)$ if $A\cap B=\emptyset$. We call ν the (finitely additive) *natural influence measure*. It is *not* supposed that ν is σ-additive. This requirement would not be consistent with the developed Boolean approach based on finite combinations of methods of measurement.

The natural influence measure ν yields in a unique way a normalized, non-negative *Radon measure* $\bar{\nu}$ defined on the Borel σ-field of Stone(BG(Φ)):

Since every clopen subset of Stone(BG(Φ)) is compact we obtain by $\nu(A) = \bar{\nu}(\bar{A} \setminus Z) \forall A \in BG(\Phi)$ a measure on the field of clopen subsets of Stone(BG(Φ)). Then following P.R.HALMOS [11], § 13, § 54, and Z.SEMADENI [33] 18.1.4(c) we obtain $\bar{\nu}$. If we interpret Stone(BG(Φ)) as a compactification of N then the above equality can be formulated as follows: $\nu(A) = \bar{\nu}(cl(A)) \forall A \in BG(\Phi)$ (cf. [21] 1.3 and [19] for more details).

1.4 CONCLUSION: We have introduced the basic concept of an idealized, *deterministic* observation process. This concept is grounded on methods of measurements having one dimensional, discrete and *finite scales*. Only *finite combinations* of such methods of measurement are considered.

Within this framework we obtained the finitely additive measure space (N,BG(Φ),ν). Here N was the set of agents and BG(Φ) was the system of coalitions, interpreted here as the system of observed groups. The finitely additive, non-negative, normalized influence measure ν was defined on the field BG(Φ). Due to the idealized observation process the triple (N,BG(Φ),ν) yielded a second one, namely (Stone(BG(Φ)),CO(Stone(BG(Φ))),$\bar{\nu}$).[1)] Here Stone(BG(Φ)), the Stone space of BG(Φ), was interpreted as the compact space of all points corresponding to relevant, possible property combinations which an agent may have. The basis CO(Stone(BG(Φ))) of clopen subsets of Stone(BG(Φ)) corresponded directly to BG(Φ) and to finite combinations of methods of measurement. Lastly the Radon measure $\bar{\nu}$ was the extension of ν onto Stone(BG(Φ)). If Φ sepa-

---

1) It would be more familiar to use the triple (Stone(BG(Φ)), $\mathcal{B}$(Stone(BG(Φ))),$\bar{\nu}$) where $\mathcal{B}$(...) denotes the Borel-σ-field. Then the triple symbolizes a probability space. However we will need only CO(Stone(BG(Φ))), and we can interpret this class. Indeed, instead of the σ-additive theory we will use the $\bar{\nu}$-Jordan content theory i.e. the notation of topological measure spaces.

rated N then the Stone space was seen as a compactification of N.

Therefore we have motivated and explained how *a discrete, finitely additive measure space yields a topological* - more precisely: a compact - *measure space*. This was the first construction necessary to change from a discrete model to a continuous one.

2. THE OBSERVATION OF A MARKET

Due to L.WALRAS [37] and G.DEBREU [9] the abstract formulation of a market is defined as follows: A (finite) *market* is a map $\mathcal{E} := E \otimes e: \{1,\dots,n\} \to P \times \mathbb{R}^n$. The set of agents is $\{1,\dots,n\}$. E assignes every agent his preference and e(i) is the commodity bundle of the i-th agent. We want to elaborate a preliminary definition of a market with infinitely many agents, and we want to describe the formal observation of such a market.

2.1 THE OBSERVATION OF ENDOWMENT MAPPINGS. (A) At first we introduce the compact metric *space* (P,δ) *of preferences* and the metric *space* (P_{mo},δ) *of monotonic preferences:*

Let $\mathbb{R}^n_+ := \{x \in \mathbb{R}^n \mid x_i \geq 0,\ 1 \leq i \leq n\}$. A *preference* is an open subset of the space $\mathbb{R}^n_+ \times \mathbb{R}^n_+$ which yields a transitive and irreflexive binary relation on $\mathbb{R}^n_+$. We denote P the set of all preferences. Since our interest is focused on the problem of pure competition we restrict ourselves to the consumption set $\mathbb{R}^n_+$.

Let $\gamma\mathbb{R}^{2n}_+ := \mathbb{R}^{2n}_+ \cup \{\infty\}$ the one point compactification of $\mathbb{R}^{2n}_+$. Since $\gamma\mathbb{R}^{2n}_+$ possess a countable basis, $\gamma\mathbb{R}^{2n}_+$ is a compact metrizable space. Due to the compactness the metric is uniquely defined up to a homeomorphism which is uniformly continuous in both directions. We identify a preference $\succ$ with the set $\bar{\succ} := \gamma\mathbb{R}^{2n}_+ \setminus \succ$. Then the Hausdorff distance derived from the metric space $\gamma\mathbb{R}_{+}^{2n}$ yields a uniformity on the set P of preferences. This uniformity is well known as the *topology of closed convergence*. Then P is a compact metric space with respect to the topology of closed convergence (for proofs and details see W.HILDENBRAND [13]).

We obtain a concrete picture of the metric on P as follows: Let S^{2n} be the unit sphere in $\mathbb{R}^{2n+1}$. Let $p:S^{2n} \longrightarrow \gamma\mathbb{R}^{2n}$ be the *spherical projection* which is a uniformly continuous homeomorphism. Then $\gamma\mathbb{R}_{+}^{2n}$ can be identified with $p^{-1}(\gamma\mathbb{R}_{+}^{2n}) \subset S^{2n}$. Let d denote the *Euclidean distance* of $\mathbb{R}^{2n+1}$. Then d restricted onto S^{2n} induces the required *Hausdorff distance* δ between any two preferences, i.e. $\delta(\succ,\succ')=\inf\{\varepsilon\in(0,\infty] \mid p^{-1}(\overline{\succ'}) \subset B_{\varepsilon}(p^{-1}(\overline{\succ}))$ and $p^{-1}(\overline{\succ}) \subset B_{\varepsilon}(p^{-1}(\overline{\succ'}))\}$ where $B_{\varepsilon}(M):=\{x\in S^{2n} \mid \exists y\in M$ with $d(x,y) <\varepsilon\}$ for $M\subset S^{2n}$ (see C.KLEIN [20] for details).

Let $\succ, \succ'$ be two preferences. One can change strongly the distance $\delta(\succ,\succ')$ if one subtracts a suitable, nowhere dense, closed subset from $\succ$ resp. from $\succ'$. Nowhere dense sets are less suitable for a measurement process. W. HILDENBRAND (cf. [13], p. 86) argues that nowhere dense sets can not be falsified. Hence we deem the closed convergence metric useful for *regular open* sets $\succ$ only. Here a set A is called regular open if int cl(A)=A.i.e. if the interior of the closure of A coincides with A.

A preference $\succ$ is called *monotonic* if $\succ$ fulfills the following condition: If y>x - i.e. $x,y\in\mathbb{R}_{+}^{n}, x_i\leq y_i \forall i$ and $x\neq y$ - then y is prefered to $x:(y,x)\in\succ$. P_{mo} denotes the set of all monotonic preferences. (P_{mo},δ) is a metric space but not a compact space i.e. P_{mo} is not a closed subset of (P,δ). Observe that an element $\succ\in P_{mo}$ is always regular open.

In the final stage of the model we will use exclusively monotonic preferences. A profile of monotonic preferences corresponds from our point of view to the natural assumption that a market can exist for *scarce* commodities only. Moreover by considering only monotonic preferences we avoid the problems which arise by using the closed convergence topology. However we will need the space (P,δ) to deduce some helpful mathematical properties.

(B) In this stage of the model we want to study the behaviour of endowment mappings.

An *endowment mapping* is a mapping $\mathcal{E}:=E\times e:N\to P\times \mathbb{R}^n_+$ such that $e(N)$ is *bounded* in $\mathbb{R}^n_+$. If $E(N)\subset P_{mo}$ then $\mathcal{E}$ is called P_{mo}-valued.

The *boundedness assumption* reflects the fact that the whole amount of a commodity in the economy is finite. However the main argument for that boundedness requirement will be deduced in the second chapter. Observe that the boundedness assumption is always fulfilled for a finite set N i.e. in the non-idealized case.

Assume that N is observed by a family Φ of methods of measurement which separates N. Then N is interpreted as a dense subset of Stone(BG (Φ)). We can answer the question of whether $\mathcal{E}$ extends continuously onto Stone(BG(Φ)) or not (see [21] for proofs and details):

We denote $\|\ \|$ the Euclidean metric on $\mathbb{R}^n_+$. Then $\delta\times\|\ \|$ is a metric on $P\times\mathbb{R}^n_+$ where δ symbolizes the Hausdorff distance on P. We obtain the *canonical sup metric* d_o on the *set* EM *of all endowment mappings* - i.e. all bounded mappings $\mathcal{E}:N\to P\times\mathbb{R}^n_+$ - as follows: $d_o(\mathcal{E},\mathcal{F}):=\sup\{\delta(E(i),F(i))+\|e(i)-f(i)\|\,|\,i\in N\}$. An endowment mapping $\mathcal{E}$ is called BG(Φ)-*simple* iff there is a finite, disjoint decomposition $\{A_i\}_{i=1}^n\subset BG(\Phi)$ of N such that $\mathcal{E}$ is constant on every A_i. Then the following results holds:

The endowment mapping $\mathcal{E}$ *possesses a unique, continuous extension* $\overline{\mathcal{E}}$ *on* Stone(BG(Φ)) *iff* $\mathcal{E}$ *can be approximated by* BG(Φ)-*simple functions with respect to the canonical sup metric* d_o.

The underlying mathematical idea of the above result is the following: Stone(BG(Φ)) is a zero-dimensional, compact space. Therefore the simple mappings on Stone(BG(Φ)) are dense in the space of continuous mappings with respect to the canonical sup metric (cf. Z.SEMADENI [33],

§ 8.2).

The above result can be formulated verbally: If BG(Φ) contains enough subsets of N - i.e. if the observation by Φ is *sufficiently precise* - then the endowment mapping $\mathcal{E}$ extends continuously to $\overline{\mathcal{E}}$ on Stone (BG(Φ)).

(C) Assume that $\mathcal{E} := E \times e : N \to P_{mo} \times \mathbb{R}^n_+$ is a P_{mo}-valued endowment mapping. We describe a natural measurement procedure which yields a countable family Φ of methods of measurement such that $\mathcal{E}$ extends continuously onto Stone(BG(Φ)) (for proofs, details, and a more general context see [21] 2.1 and 2.2).

(i) Let $r>0$ and $m \in \mathbb{N} := \{1,2,3...\}$. If $i \in \{0,1,2,...,m-1\}$ then $S_i := [ir/m,(i+1)r/m)$. Let $s_i := ir/m + r/2m$ if $0<i<m$ and let $s_0=0$. We call $\alpha = \alpha(r,m) := (\{S_i\}_{0\leq i<m}, \{s_i\}_{0\leq i<m}, r)$ a *scale observation of* $\mathbb{R}_+$. We denote $gr(\alpha) := \{s_i\}_{0\leq i<m}$ the *grid*, $r(\alpha) := r$ the *range* and $grc(\alpha) := r/m$ the *grid constant* with respect to α. We mention that we defined $s_0=0$ to avoid technical difficulties later on at the boundary of $\mathbb{R}_+$ in $\mathbb{R}$. From a canonical product construction we obtain a *scale observation* α^n *of* $\mathbb{R}^n_+$. Observe that $gr(\alpha^n(r,m))$ possesses m^n grid points and that $r(\alpha)=r(\alpha^n)$, $grc(\alpha)=grc(\alpha^n)$. A sequence of scale observations $\{\alpha_i\}_{i\in\mathbb{N}}$ of $\mathbb{R}_+$ (resp. of $\mathbb{R}^n_+$) is called *growing* iff $r(\alpha_i)\to\infty$ and $grc(\alpha_i)\to 0$. We prefer here the above specialized notation of a grid to avoid technical details which are less important for the principal reasoning. The essential characteristics of the applied grids are finite ranges and finitely many grid points contained in the interiors of the related scale intervals.

(ii) At first we describe the measurement procedure with respect to the commodity allocation e. Let $f: N \to \mathbb{R}_+$ be a bounded function and let $\alpha(r,m)$ be a scale observation of $\mathbb{R}_+$. We define $f_\alpha : N \to \mathbb{R}_+$ as follows: $f_\alpha(n) := s_i$ if $f(n) \in S_i$ and $f_\alpha(n) := s_{m-1}$ if $f(n) \geq r$. Then f_α is called the *observed function* with respect to f and α.

A sequence $\{\alpha_i\}_{i\in\mathbb{N}}$ of scale observations is called *approximating* with respect to f if $\sup\{|f(n)-f_{\alpha_i}(n)| \mid n\in\mathbb{N}\}\to 0$ for $i\to\infty$. Remark that *a growing sequence of scale observations is approximating with respect to every bounded function from* N *into* $\mathbb{R}_+$.

The bounded function f and the scale observation $\alpha(r,m)$ of $\mathbb{R}^n_+$ yield *finitely* many methods of measurement $\{\varphi_i\}_{0\leq i\leq m}$ as follows: $G(i)=\{n\in N \mid \varphi_i(n)=1\}=\{n\in N \mid f(n)\in S_i\}$ if $0\leq i<m-1$ and $G(i)=\{n\in N \mid f(n)\in S_i$ or $f(n)\geq r\}$ if $i=m-1$. Let $\{\alpha_j\}_{j\in\mathbb{N}}=\{\alpha_j(r_j,m_j)\}_{j\in\mathbb{N}}$ a growing sequence of scale observations of $\mathbb{R}^n_+$. Then $\{\alpha_j\}_{j\in\mathbb{N}}$ yields analogously an at most *countable* family $\Phi=\{\varphi_{ij}\}_{0\leq i\leq m_j, j\in\mathbb{N}}$ of methods of measurement. The generated Boolean algebra BG(Φ) is countable too. Every observed function f_{α_j} with respect to α_j, $j\in\mathbb{N}$ and f is a BG(Φ)-simple function, and $f_{\alpha_j}\to f$ with respect to the sup norm over N. Therefore f extends continuously onto Stone (BG(Φ)).

We consider now a commodity allocation $\hat{e}=(e_1,\ldots,e_n):N\to\mathbb{R}^n_+$ consisting of n bounded mappings $e_i:N\to\mathbb{R}_+$, $1\leq i\leq n$. Obviously we obtain then the analogous result with respect to a growing sequence of scale observations of $\mathbb{R}^n_+$, i.e. if the family of methods of measurement Φ is generated analogously as above, *then* Φ *is countable and e extends continuously onto* Stone(BG(Φ)).

(iii) We want to explain the measurement procedure with respect to a P_{mo}-valued profile $E:N\to P_{mo}$. This procedure is similar constructed as that of part (ii). The main problem and difference to (ii) is the definition of an observed profile:

Let $n_o\in N$ be an agent with $E(n_o)=\succ\in P_{mo}$. Let $\alpha=\alpha(r,m)=(\{S_i\}_{0\leq i<m};\{s_i\}_{0\leq i<m};r)$ a scale observation of $\mathbb{R}_+$ and let $\alpha^n=\alpha^n(r,m)$ the generated scale observation of $\mathbb{R}^n_+$. Assume that $s=(s_{i1},\ldots,s_{in})$ and $s'=(s_{j1},\ldots,s_{jn})$, $0\leq ik, jk<m$ are two grid points of $\alpha^n(r,m)$. Then with a single one bit question we obtain knowledge whether (s,s') is contained in $\succ$ or

not. Hence with *finitely many questions* we obtain all grid points of $\alpha^n(r,m)$ contained in $\succ$. We denote $A(\succ,\alpha^n) := \bigcup_{o \leq ik, jk < m} \{(S_{i1} \times \ldots \times S_{in}) \times (S_{j1} \ldots S_{jn}) | ((s_{i1},\ldots,s_{in}),(s_{j1},\ldots,s_{jn})) \in \succ\}$. We define $\succ(\alpha^n) :=$ int cl$(A(\succ,\alpha^n))$ i.e. $\succ(\alpha^n)$ is the interior of the closure of $A(\succ,\alpha^n)$ in $\mathbb{R}^n_+ \times \mathbb{R}^n_+$. *Then* $\succ(\alpha^n)$ *is an element of* P (cf. [21] prop. 8). Moreover $\succ(\alpha^n)$ is regular open by definition. We call $\succ(\alpha^n)$ *the observed preference* with respect to $\succ$ and α^n.

Let $E:N \rightarrow P_{mo}$ be a P_{mo}-valued profile. Then *finitely many methods of measurement* yield analogously as above the *observed profile* $E_\alpha n:N \rightarrow P$ with respect to α^n. Remember that $E_\alpha n(n)$ is regular open for every $n \in N$. Therefore we can apply the closed convergence metric without difficulties for the interpretation.

A sequence $\{\alpha^n_i\}_{i \in \mathbb{N}}$ of scale observations of $\mathbb{R}^n_+$ is called *approximating* with respect to the P_{mo}-valued profile E if $\sup\{\delta(E(n),E_{\alpha_i}n(n)) | n \in N\} \rightarrow 0$ for $i \rightarrow \infty$. *A growing sequence of scale observations of* $\mathbb{R}^n_+$ *is approximating with respect to every* P_{mo}*-valued profile*. This is a consequence of the following result (cf. [21] prop. 9 for a proof):

<u>PROPOSITION I.1.</u> Let $\{\alpha^n_i\}_{i \in \mathbb{N}}$ be a growing sequence of scale observations of $\mathbb{R}^n_+$. Let $\varepsilon > 0$. Then there is an $i_o \in \mathbb{N}$ such that $\delta(\succ, \succ_{\alpha^n_i}) < \varepsilon$ for every $\succ \in P_{mo}$ and every $i \geq i_o$.

A growing sequence of scale observations of $\mathbb{R}^n_+$ delivers, in analogy to (ii), *countably many* methods of measurement with respect to the P_{mo}-valued profile E. Therefore we obtain a countable Boolean algebra $BG(\Phi)$, and *every observed profile* E_{α_i} *is* $BG(\Phi)$*-simple*. This implies that E *extends continuously onto* Stone $BG(\Phi)$, i.e. there is a unique continuous mapping $\bar{E}:\text{Stone}(BG(\Phi)) \rightarrow P$ with $\bar{E}|_N \equiv E$.

(iv) Assume that the idealized set N of agents is *countable*. Then it is obvious that a *countable system* Φ of methods of measurement exists which *separates* N.

Let us summarize our results: Assume that $\mathcal{E}:=E\times e:N\to P_{mo}\times \mathbb{R}_+^n$ is an endowment mapping and that N is countable. Then there is an at most *countable* system Φ of methods of measurement such that Φ separates N and such that $\mathcal{E}$ extends *uniquely and continuously* onto Stone(BG(Φ)), i.e. there is a continuous mapping $\bar{\mathcal{E}}:=\bar{E}\times\bar{e}:\text{Stone}(BG(\Phi))\to P\times\mathbb{R}_+^n$ with $\bar{\mathcal{E}}|_N\equiv\mathcal{E}$. We call $\bar{\mathcal{E}}$ the *observed endowment mapping* with respect to Φ and $\mathcal{E}$.

The countable system Φ of methods of measurement was deduced by a naturally idealized measurement process: We put on $\mathbb{R}_+^n$ finitely many grid points. The finitely many questions related to the grid points deliver an endowment mapping which corresponds to an observation with a finite precision. A sequence of such grid systems with growing fineness yields a sequence of observations with growing precision. The resulting limit observation is our observed endowment mapping $\bar{\mathcal{E}}$.

2.2 <u>THE OBSERVATION OF PROPERTIES</u>. In the foregoing part we could not guarantee that $\bar{E}(n)\subset P_{mo}$ in the case of $E(N)\subset P_{mo}$. With respect to a countably infinite set N of agents we want to solve this problem in a principal discussion on the qualifications of our observation procedure.

For this purpose we consider the example of part 1.2 where $N=\mathbb{N}$ and where Stone(BG(Φ)) was the one point compactification $\gamma\mathbb{N}$ of $\mathbb{N}$. Let $e:\mathbb{N}\to\mathbb{R}_+$ be defined by $e(n)=1-1/n$ for $n\in\mathbb{N}$. Then e extends continuously onto $\gamma\mathbb{N}$ and $\bar{e}(\infty)=1$. Hence the statement "The function is everywhere less than one" is true for e and false for $\bar{e}$. One recognizes that e possesses a special property which $\bar{e}$ fails to maintain.

The above described situation can not occur if N is finite or if we describe e(n) by dollars and cents without the idealized mathematical object $\mathbb{R}_+$. The above situation is an idealized mathematical one because it depends on the behaviour of e at infinity. We want to use the mathematical framework as a help for our analysis, but we are not merely interested in problems arising only due to the mathematical idealization. We argue that the relevant properties of our markets have to be of a somewhat finite quality. The meaning of what we called finite quality shall be elaborated now:

Let us study the above problem in a different way: Assume that $\alpha:=(\{S_i\}_{o\leq i<m};\{s_i\}_{o\leq i<m};r)$ is a scale observation of $\mathbb{R}_+$. Then e_α is less than one on $\mathbb{N}$ only if α is chosen in a special manner. Suppose that $\{\alpha_j\}_{j\in\mathbb{N}}$ is a growing sequence of scale observations of $\mathbb{R}_+$. Then we can not guarantee that any observed function possesses the above mentioned property, unless we choose (r_j,m_j) in a very special way.

Assume that we are studying a relevant characteristic of the endowment mapping e with respect to any growing sequence of scale observations of $\mathbb{R}_+$. We deem an observation by a growing sequence to correspond to measurements with an *arbitrarily fine, but always finite, precision.* Consequently we argue that there shall be a $n_o\in\mathbb{N}$ such that e_{α_j} possesses the relevant characteristic for all $j\geq n_o$.

We formulate the above deduced requirement on our model construction in a principal manner: (A4) *A property of a given endowment mapping is called observable iff there exists a fixed, finite precision guaranting that every more precise measurement verifies that property of the endowment mapping*. Roughly speaking: A property is observable iff it can always be verified by measuring with a fine enough, finite precision. This was the concrete meaning of what we called a finite quality of properties.

The above axiom (A4) aims *not* at the measurement in reality. Indeed with respect to real markets - i.e. finite discrete markets - the axiom is always fulfilled. Instead of that (A4) is a *rule* concerning a reasonable idealization and a suitable, mathematical modelling. This rule will enable us to idealize *economic* assumptions on our model in a reasonable, mathematical manner. From our viewpoint, (A4) can be seen as a straight forward consequence of the earlier axioms (A1)-(A3).

In the following we elaborate mathematically types of properties for endowment mappings which are consistent with respect to our model (for details and proofs see [21] 1.5 and 2.2).

A mapping $p:\mathbb{P}(P \times \mathbb{R}^n_+) \rightarrow \{0,1\}$ is called an *endowment property*. Here $\mathbb{P}(P \times \mathbb{R}^n_+)$ denotes the power set of $P \times \mathbb{R}^n_+$. A mapping $\mathcal{E}:N \rightarrow P \times \mathbb{R}^n_+$ *possesses* the endowment property p iff $p(\mathcal{E}(N))=1$. $\mathcal{E}$ is called p-*stable* if there is an $\varepsilon>0$ such that every mapping $\mathcal{T}:N \rightarrow P \times \mathbb{R}^n_+$ with $\mathcal{T}(N) \subset B_\varepsilon(\mathcal{E}(N))$ implies $p(\mathcal{T}(N))=1$. Here $B_\varepsilon(\mathcal{E}(N))$ denotes the open ε-ball around $\mathcal{E}(N)$. The endowment property p is called *stable* (or continuous) if $p(\mathcal{E}(N))=1$ implies that $\mathcal{E}$ is p-stable. The endowment property p is called *scaling invariant* if the following holds: Let $\{\alpha_i\}_{i\in\mathbb{N}}$ be a growing sequence of scale observations of $\mathbb{R}_+$. Assume that $p(\mathcal{E}(N))=1$ and that $\{\alpha_i^n\}_{i\in\mathbb{N}}$ is approximating with respect to $\mathcal{E}$. Then there is an $i_o\in\mathbb{N}$ such that the observed endowment mappings $\mathcal{E}_{\alpha_i^n}$ possess the endowment property p for all $i \geq i_o$. (The definitions are applied analogously for a fixed subset $S \subset N$.)

<u>REMARK I.2.</u> Let p be an endowment property and let $\mathcal{E}:N \rightarrow P_{mo} \times \mathbb{R}^n_+$ be an endowment mapping. Then the following holds: (i) If $\mathcal{E}$ is p-stable then the extension $\overline{\mathcal{E}}$ of $\mathcal{E}$ onto Stone(BG(Φ)) is p-stable (supposing that $\overline{\mathcal{E}}$ exists). (ii) If p is stable, then $p(\mathcal{E}(N))=1$ iff $p(\overline{\mathcal{E}}(\text{Stone}(BG(\Phi))))=1$ (supposing that $\overline{\mathcal{E}}$ exists). (iii) If p is stable, then p is scaling invariant.

<u>EXAMPLE I.1.</u> (a) We define $pa:P \times \mathbb{R}^n_+ \rightarrow \{0,1\}$ as follows: $pa(A)=pa(A_1 \times A_2)=1$ iff there is an $\varepsilon>0$ such that $\max\{a_i | 1\leq i\leq n\}>\varepsilon$ $\forall a=(a_1,\dots,a_n)\in A_2 \subset \mathbb{R}^n_+$. Then pa is a stable endowment property. (b) Let $x,y\in\mathbb{R}^n_+$. Then we define $pb_{x,y}:P \times \mathbb{R}^n_+ \rightarrow \{0,1\}$ as follows: $pb_{x,y}(A)=pb_{x,y}(A_1 \times A_2)=1$ iff there is an open subset U of $\mathbb{R}^{2n}_+$ such that $(x,y)\in U\subset \succ_a$ for all $\succ_a \in A_1$. Then $pb_{x,y}$ is a stable endowment property. (c) The statements "Every

agent has a positive first endowment" and "Every agent has a monotone preference" yield no scaling invariant endowment properties.

The above results are applied now to obtain a suitable abstract formulation of the economic assumption of monotonicity, i.e. of the assumption that the commodities are scarce.

Let $\mathcal{E}: N \to P_{mo} \times \mathbb{R}^n_+$ be an endowment mapping. If $x,y \in \mathbb{R}^n_+$, $x>y$, then *the monotonicity of* $\mathcal{E}$ *in* (x,y) *is called observable* iff $pb_{x,y}(\mathcal{E}(N))=1$, i.e. iff there is an open set U, $(x,y) \in U \subset \mathbb{R}^{2n}_+$ with $U \subset \succ$ for all $\succ \in \mathcal{E}(N)$. Due to the scaling invariance of $pb_{x,y}$ the above definition is consistent with the measurement principle (A4).

An *endowment mapping* $\mathcal{E}: N \to P_{mo} \times \mathbb{R}^n_+$ *is called monotonic* iff the monotonicity of $\mathcal{E}$ is observable in every pair $(x,y) \in \mathbb{R}^{2n}_+$ with $x>y$.

Neither the monotonicity of a single preference nor the monotonicity of an endowment mapping can be observed in the sense of the principle (A4). The above definition is a *model consistent*, mathematical formulation of the *economic assumption* that monotonicity holds. If N is finite then every mapping $\mathcal{E}: N \to P_{mo} \times \mathbb{R}^n_+$ is monotonic. Hence the above definition is only a careful analysis of the idealized case. This improves the following result:

Let $\mathcal{E}: N \to P_{mo} \times \mathbb{R}^n_+$ be an endowment mapping such that the extension $\overline{\mathcal{E}}: \mathrm{Stone}(BG(\Phi)) \to P \times \mathbb{R}^n_+$ exists. Then *the range of* $\overline{\mathcal{E}}$ *is contained in* $P_{mo} \times \mathbb{R}^n_+$ *iff* $\mathcal{E}$ *is monotonic* (cf. [20] for an explicit proof).

We are prepared now for the definition of a market mapping with respect to the deterministic framework of this first chapter. A *market mapping* is a bounded, monotonic endowment mapping $\mathcal{E}: N \to P_{mo} \times \mathbb{R}^n_+$. Then $\overline{\mathcal{E}}: \mathrm{Stone}(BG(\Phi)) \to P_{mo} \times \mathbb{R}^n_+$ is called the *observed market mapping.* (Here Φ is e.g. obtained as in part 2.1 to guarantee that $\overline{\mathcal{E}}$ exists.)

We change our viewpoint: Let K be a compact, zero-dimensional space, i.e. a Stone space. Let N be a countably infinite set densly contained in K. Let $\bar{\mathcal{E}}:K \to P_{mo} \times \mathbb{R}^n_+$ be a continuous mapping. *The restriction* $\mathcal{E}$ *of* $\bar{\mathcal{E}}$ *onto* N *is a market mapping.* Hence from the abstract viewpoint the situation turns out to be rather simple.

We want to mention one point: Our observation procedure is based on growing sequences of grids and the deduced "rectangular-preferences". Therefore it should be reasonable to construct the space of preferences as a completion of the set of "rectangular-preferences". Then, we could avoid the critics with respect to the topology of closed convergence. Although we deem such a completion to be possible there are two reasons why we have not presented such an attempt: (1) The topology of closed convergence is well known. (2) The results with respect to P_{mo} remain nearly the same. We are focused here on P_{mo}.

2.3 THE OBSERVATION OF MEASURES. We begin here to close the gap between the abstract, finitely additive natural influence measure and the real situation which is characterized by finitely many agents and the Laplace probability. The final concretization of the unique natural influence measure is reserved for the second chapter.

We start by considering a finite set N of agents. Let $\Psi:\{1,...,n\} \xrightarrow{\simeq} N$ be any numeration of N. Assume that $f:N \to \mathbb{R}$ is a function. Then the integral with respect to the Laplace probability l_n is given by $\int_N f dl_n = (1/n)\sum_{i=1}^n f(\Psi(i))$. If $f=\chi(A)$, where $\chi(A)$ is the characteristic function of a group $A \subset N$, then we obtain the l_n-value of A, i.e. $l_n(A) = \text{card}(A)/\text{card}(N)$. Hence in the finite case we obtain *all* information concerning the natural influence measure and integrals by summations over the agents.

We rewrite the above formula as follows: $(1/n)\sum_{i=1}^n f(\Psi(i)) = \lim_{m\to n}(1/m)\sum_{i=1}^m f(\Psi(i))$. Assume that the numeration is carefully selected – i.e. randomly. Assume that m is interpreted as the size of a random sample and that f fulfills sufficient requirements. Then the above "limit" corresponds to practical statistics. This interpretation of the above sum is a motivation of the following, idealized summability approach. It is not our goal to interpret the model from the point of view of statistics.

Assume that N is an idealized, countably infinite set of agents. We considered a market mapping $\mathcal{E}$ and deduced the mathematical counterpart $\overline{\mathcal{E}}$ on the related Stone space. In the idealized case the natural influence of every agent $n \in N$ shall be zero. Therefore it seems natural to neglect the agents and to calculate integrals on the related Stone space. However the mapping $\overline{\mathcal{E}}$ was a *technical* help only. *All* information in our model has to be deduced from the agents' characteristics $\mathcal{E}$. Therefore, we conclude that the following principle has to be valid:
(A5) *Results obtained by integration are model consistent iff they can be deduced equivalently on the set* N *of agents.*

The above stated requirement can be fulfilled in a natural way by the theory of uniformly distributed sequences (cf. the monographs [14] of E.HLAWKA and [24] of L.KUIPERS and H.NIEDERREITER for details and proofs):

Let K be a compact, metrizable space endowed with a normalized, non-negative Radon measure μ. Let $\{x_i\}_{i\in\mathbb{N}}$ be a sequence in K. Then $\{x_i\}_{i\in\mathbb{N}}$ is called μ-*uniformly distributed* iff

(1) $$\lim_{n\to\infty} \frac{1}{n} \sum_{i=1}^{n} f(x_i) = \int_K f d\mu$$

holds for every continuous real valued function f on K. (i) We interpret a sequence $x := \{x_i\}_{i\in\mathbb{N}} \subset K$ as an element of the countable product $K_\infty := \Pi_1^\infty K$. With respect to the canonical product measure $\mu_\infty := \Pi_1^\infty \mu$ almost all sequences x are μ-uniformly distributed. (ii) Assume that the support $\mathrm{supp}(\mu) := \{x \in K \mid \mu(U(x)) > 0$ for every open neighbourhood $U(x)$ of $x\}$ of μ

possesses no isolated point and that S is a countable, dense subset of K. Then S can be enumerated in such a way that S is μ-uniformly distributed.

We verify the above assumptions in our situation: Since Φ can be assumed to be countable it follows that Stone(BG(Φ)) is a compact, metrizable space. The set N of agents is a countable, dense subset of Stone (BG(Φ)). In the sequel we will recognize that supp($\bar{\nu}$) has no isolated point provided that pure competition holds. Then we obtain the following consequence of (ii): In our sitation there exists an enumeration Ψ: $\mathbb{N} \xrightarrow{\simeq} N$ such that $\{\Psi(i)\}_{i\in\mathbb{N}}$ is $\bar{\nu}$-uniformly distributed in Stone(BG(Φ)). If A is a group - i.e. $A \in BG(\Phi)$ - then $\chi(cl(A))$ is a continuous function on Stone(BG(Φ)). As $\nu(A)=\bar{\nu}(cl(A))=\int\chi(cl(A))d\bar{\nu}$ we can calculate $\nu,\bar{\nu}$, and integrals of continuous functions as in formula (1) - i.e. equivalently on N or on Stone(BG(Φ)). Therefore the above stated requirement on our observation model is fulfilled.

Observe the following interpretation of (i): If Ψ is chosen by random, then Ψ yields a $\bar{\nu}$-uniformly distributed sequence with probability one. This seems to us to be an appropriate counterpart to the above mentioned motivation by statistics.

The above consideration implies the following consequence: *From now on the notation* $N \equiv \mathbb{N}$ *will be used.* The Stone space Stone(BG(Φ)) is interpreted as a compactification of $\mathbb{N}$. *We assume that integrals of continuous functions can be calculated analogously as in formula* (1); i.e. an integral is seen as a limit of average means.

A property of the natural influence measure shall be analyzed in the sequel: We are faced with the question about what it means that pure competition is observed. The answer to this question is *not* obvious for we have no finite counterpart of such a definition. Hence we want to motivate and elaborate a definition of pure competition which is consistent with respect to our observation model.

For this reason we consider once again the finite situation: Let $N=\{1,\dots,n\}$ and let l_n be the Laplace probability where $l_n(i)=1/n$. We accept l_n as the natural influence measure on N and interpret l_n as a finite approximation of pure competition. Then we distinguish two different ways of an idealization:

(α) We pick out any agent i and we recognize that $l_n(i) \longrightarrow 0$ for $n \to \infty$. Hence we obtain the natural requirement that $\nu(i)=0$ has to be fulfilled in the idealized case for every agent $i \in \mathbb{N}$.

Let us discuss the consequences which we obtain by accepting the above requirement as a definition of pure competition: This possible definition is based on single agents and not on groups. Our model is derived from the behaviour of the agents. However, in general we can not identify a single agent since this would imply to ask countably infinitely many questions. A single agent remains anonymous in general. We can identify groups only. Therefore the above possible definition is not consistent with respect to our observation principles.

Moreover a lack of stability is inherent to this possible definition: Assume that $\nu(i)=0$ for every agent $i \in \mathbb{N}$. Furthermore assume that there is an $x_o \in \mathrm{Stone}(BG(\Phi))$, $x_o \notin \mathbb{N}$ with $\bar{\nu}(x_o)>0$. Then x_o is a possible property combination. Therefore we can declare x_o to be an agent - i.e. a realized property combination - without changing the resulting observation model. Consequently the observation model would yield no help to decide whether pure competition holds or not. Due to these two reasons we are forced to abandon this possibility to define pure competition.

(β) We describe a second way to idealize the Laplace probability l_n: If $N=\{1,\dots,n\}$ then for every $\varepsilon \geq 1/n$ there are finitely many subsets $A_j \subset N$, $1 \leq j \leq m(\varepsilon)$ with $N \subset \bigcup_{j=1}^{m(\varepsilon)} A_j$ and $l_n(A_j) \leq \varepsilon$ for $1 \leq j \leq m(\varepsilon)$. With $N=\mathbb{N}$ and $n \to \infty$ we obtain an idealized condition. Since it is the appropriate one, we formulate:

DEFINITION I.3. Let N be a countable set of agents which is observed by a system Φ of methods of measurement. Then *pure competition holds* iff for every $\varepsilon>0$ there are finitely many groups $A_j \in BG(\Phi)$, $1 \leq j \leq m(\varepsilon)$ with $N \subset \bigcup_{j=1}^{m(\varepsilon)} A_j$ and $\nu(A_j)<\varepsilon$ for $1 \leq j \leq m(\varepsilon)$. Here ν denotes the natural influence measure.

Observe that the restriction $A_j \in BG(\Phi)$ is necessary because ν is defined on $BG(\Phi)$ only. Moreover, due to this restriction, the definition is consistent with our observation principles which are based on groups. Considering any zero sequence $\varepsilon_n \rightarrow 0$, $\varepsilon_n>0$ we recognize that a countable Boolean algebra $BG(\Phi)$ suffices to fulfill the above condition. One easily deduces that $\nu(i)=0$ for all $i \in N$ is fulfilled in the case of pure competition. The following proposition solves the above discussed stability problem (for a proof see [19]):

PROPOSITION I.4. Let N be a countable set of agents which is observed by a system Φ of methods of measurement. Then pure competition holds iff the extended natural influence measure $\bar{\nu}$ is atomless (i.e. $\bar{\nu}(x)=0$ for every $x \in \mathrm{Stone}(BG(\Phi))$).

The property of pure competition can be examined on N by ν or, equivalently, on Stone $(BG(\Phi))$ by $\bar{\nu}$. Therefore we deem the above definition to be suitable for our model. It is obvious that the support of an atomless measure has no isolated points. This implies that the summability approach from the beginning of this section can be applied in the case of pure competition.

We will describe in chapter II the calculation of ν as a limit of the l_n-measures. Moreover we will give there a concretization of Φ as well as a concretization of the above mentioned groups A_j.

We scetch verbally the results of this part once more: An enumeration of the set N of agents, similar to a random enumeration, yields a picture of the natural influence measure by a limit of average means. Results obtained by integration are consistent with the observation model iff they can be deduced equivalently on N or on the Stone space related to the observation. Since the agents cannot be identified perfectly in the Stone space this requirement yields a limitation for the properties of the natural influence measure which are consistent with the observation principles. In particular, pure competition holds iff the whole natural influence can always be distributed to finitely many, disjoint groups having group influences less than a prescribed bound.

2.4 CONCLUSION. In the first paragraph we introduced the three basic tools of the observation concept: We applied *deterministic* methods of measurement having *finite scales*, and we allowed only *finite combinations* of these methods. In this second paragraph logical consequences of that grounding were developed:

We introduced a natural *approximation approach* which was merely based on a systematic refinement and enlarging of the applied scales of measurement. Due to this approximation we deduced that a *countable* system Φ of methods of measurement is sufficient to guarantee that: (a) Φ separates the countable set N of agents such that Stone(BG(Φ)) is a metrizable compactification of N; (b) the considered, bounded endowment mapping $\mathcal{E}: N \to P_{mo} \times \mathbb{R}^n_+$ extends *continuously* and uniquely onto the related Stone space to $\overline{\mathcal{E}}: \text{Stone}(BG(\Phi)) \to P \times \mathbb{R}^n_+$.

A straight forward consequence of the approximation approach was a logical analysis of market properties concerning their consistency with the underlying observation concept. The approximation was interpreted as measurements with an arbitrarily fine, but always finite precision. Consequently, we had to conclude that *a market property is model consistent iff it can be verified by measuring with a fine enough, finite precision*. Due to the idealized character of the model, this statement was considered as a *rule* to model abstractly economic properties. Based on this principal rule we concluded that monotonicity holds in a market iff *the monotonicity can always be verified in arbitrarily chosen, finitely many commodity pairs with respect to a finite measurement precision.* In the abstract language this definition was translated as follows: $\mathcal{E}$: $N \rightarrow P_{mo} \times \mathbb{R}_+^n$ is *monotonic* iff the continuous extension $\overline{\mathcal{E}}$ onto the related Stone space is $P_{mo} \times \mathbb{R}_+^n$ valued.

As the system Φ of methods of measurement could be supposed to be countable we derived an approximation approach for the natural influence measure: The countably infinite set N of agents was interpreted as a uniformly distributed sequence in the related Stone space. This corresponded to the assumption that the law of large numbers is valid resp. that $N \cong \mathbb{N}$ is enumerated as by random. Then *integrals* of continuous functions *could be calculated by limits of average means over the agents.* In particular, this approximation concept was consistent with the following logically deduced requirement on the modelling: *Results obtained by integration on the related Stone space are model consistent iff they can be deduced equivalently from the characteristics of the real agents only.* An analysis within the approximation concept allowed us to install and to justify carefully the model consistent definition of pure competition: *Pure competition holds iff the whole natural influence can always be distributed to finitely many, disjoint groups having group influences less than a prescribed bound.*

We summerize that the above developed approximation tool enabled us to reach two major results: Firstly, we could elaborate and justify the model consistent definitions of *monotonicity* and of *pure competition.* This was a reflection of the continuous model onto the original, discrete model. Secondly, we obtained a (deterministic) *market with pure competition* as the restriction of $\overline{\mathcal{E}}$ onto $\mathbb{N}$. Here $\overline{\mathcal{E}}$: (Stone(BG(Φ)),CO(Stone (BG(Φ))),$\bar{\nu}$) $\rightarrow P_{mo} \times \mathbb{R}_+^n$ is continuous, $\bar{\nu}$ is atomless, and Stone(BG(Φ)) is a compact, zerodimensional, metrizable compactification of $\mathbb{N}$. $\overline{\mathcal{E}}$ symbolized the *observed* (deterministic) market with pure competition. The interpretation of the Stone space tool remained the same as before.

In the first paragraph the observation concept was still based on the use of single methods of measurement. Due to the developed approximation tool we recognize that the entire observation satisfies the requirements of a discretized, but complete scanning. This *picture of a scanning* seems to us to reflect in a characteristic way the nature and the possibilities of a microeconomic analysis within the case of pure competition. Although we will simplify and concretize in the following the model by canonical, mathematical principles, we deem the above developed approach as a major *microeconomic* - not mathematic - *cornerstone* of the whole concept.

CHAPTER II. THE REPLICA MODEL

In the first chapter we described the general principles of observation used in our modelling. Strictly based on these principles we installed a deterministic, continuous model of an observed market with pure competition. Hence until this moment we took advantage mainly of the interaction between reality and idealization. In the second chapter we apply purely mathematical principles, common to all completion procedures, to transform the general model into a simplified and concrete one. The probabilistic character is then introduced into the modelling process by the Jordan content-Riemann integration theory. The resulting probabilistic model will allow us a presentation of the model in an elementary way.

A comprehensive introduction into the generalized Jordan content-Riemann integration theory can be found in the appendix.

3. THE DETERMINISTIC REPLICA MODEL

In the first chapter a preliminary definition of a market with pure competition was introduced. Although we have motivated explicitly this definition, a market with pure competition remains an idealized, mathematical object. We want to single out of the whole class of these objects a subclass which is sufficient for our purposes. The selection rule can be sketched as follows:

In the sequel we will recognize that a market with finitely many agents yields an idealized market on $\mathbb{N}$ by an infinite replication. Hence we can interpret finite markets as idealized markets. The idealized market mappings are a metric space with respect to the sup metric d_o. We will allow only those market mappings which are accumulation points of finite markets. Hence we will interpret our model as a *mathematical completion* of the class of finite markets.

The risk to create purely mathematical problems without a relation to reality is always inherent to idealizations formulated in the mathematical language. We think that the explicit motivation of the observation process combined with the mathematical completion principle will reduce this risk as much as possible.

3.1 <u>FINITE MARKETS</u>. Let $N:=\{1,\ldots,n\}$ an enumerated finite set of agents endowed with the Laplace probability l_n. The Laplace probability was defined on the power set $\mathbb{P}(N)$ of N by $l_n(i)=1/n$. Let $\mathcal{F}:N\longrightarrow P_{mo}\times\mathbb{R}^n_+$ be a finite market mapping. Then we obtain the *finite market* $\mathcal{F}:(N,\mathbb{P}(N),l_n)\longrightarrow P_{mo}\times\mathbb{R}^n_+$. We want to associate with $\mathcal{F}$ an idealized market $\mathcal{E}$ on $\mathbb{N}$. For this purpose we introduce some notations:

We denoted $\mathbb{N}=\{1,2,3\ldots\}$. If $a\in\mathbb{N}$, $b\in\mathbb{N}\cup\{0\}$ and $b<a$ then we define $M[a,b](\mathbb{N}):=\{ar-b\mid r\in\mathbb{N}\}$, e.g. $M[2,0](\mathbb{N})$ are the even numbers and $M[2,1]$ are the odd ones.

We define $\mathcal{E}:\mathbb{N}\longrightarrow P_{mo}\times\mathbb{R}^n_+$ as follows: $\mathcal{F}(n-i):=\mathcal{E}(r)$ for all $r\in M[n,i](\mathbb{N})$ where $i\in\mathbb{N}\cup\{0\}$ and $i<n$. Obviously $\mathcal{E}$ is a countably infinite *replication* of $\mathcal{F}$. We denote BG(n) as *the Boolean algebra generated by the class* $\{M[n,i](\mathbb{N})\mid i\in\mathbb{N}\cup\{0\}$ and $i<n\}$. Then BG(n) is a Boolean subalgebra of $\mathbb{P}(\mathbb{N})$ and BG(n) is canonically Boolean isomorphic to $\mathbb{P}(N)$. We define on BG(n) the natural influence measure by the *density* d:

$$d(A)=:\lim_{k\to\infty}\frac{1}{k}\sum_{r=1}^{k}\chi_A(r).$$

Here $A\in BG(n)$ and χ_A denotes the characteristic function of A, i.e. $\chi_A(r)=1$ iff $r\in A$. Observe that d on BG(n) corresponds canonically to the Laplace probability l_n on $\mathbb{P}(N)$ with respect to the canonical Boolean isomorphism between BG(n) and $\mathbb{P}(N)$.

We conclude that the finite market $\mathcal{F}:(N,\mathbb{P}(N),l_n)\longrightarrow P_{mo}\times\mathbb{R}^n_+$ yields by a countably infinite replication the idealized *replicated market* $\mathcal{E}$: $(\mathbb{N}, BG(n), d) \longrightarrow P_{mo} \times \mathbb{R}^n_+$. $\mathcal{E}$ and $\tilde{\mathcal{E}}$ are canonically equivalent. The facts that BG(n) fails to separate $\mathbb{N}$ and that $\mathcal{E}$ is not a market with pure competition are necessary consequences of this equivalence. Nevertheless we can state that *a finite market can be interpreted as an idealized market* with the help of a countably infinite replication.

3.2 THE BOOLEAN ALGEBRA OF PERIODIC SETS. In the idealized model the natural numbers $\mathbb{N}$ are considered as the set of agents.

We introduce and study now the Boolean algebra of all observable groups of the idealized, concrete model. Due to the interpretation of the concrete model as a completion we introduce that Boolean algebra here in a somewhat constructive manner. This does not mean that our interpretation based on methods of measurement is cancelled. We will discuss this interpretation when enough properties of the concrete Boolean algebra of observable groups are elaborated.

We consider the class $PER:=\{\{nr-i \mid r\in\mathbb{N}\} \mid n\in\mathbb{N},\ i\in\mathbb{N}\cup\{0\},\ i<n\}=\{M[n,i](\mathbb{N}) \mid n\in\mathbb{N},\ i\in\mathbb{N}\cup\{0\},\ i<n\}$ to be contained in the power set $\mathbb{P}(\mathbb{N})$ of $\mathbb{N}$. The Boolean algebra R_o generated by PER is called the *proper Replica algebra*. Due to the definition, R_o is the smallest Boolean subalgebra of $\mathbb{P}(\mathbb{N})$ which contains BG(n) for every $n\in\mathbb{N}$. *We declare the proper Replica algebra* R_o *to be the Boolean algebra of all observable groups* of the idealized, concrete model. Since R_o is the Boolean algebra generated by all BG(n), $n\in\mathbb{N}$, this definition is appropriate for our goal to construct a completion of all finite markets. One recognizes that R_o separates $\mathbb{N}$.

A set $A\subset\mathbb{N}$ is called *purely periodic* iff there is an $n_o\in\mathbb{N}$ such that $A\subset BG(n_o)$. If A is purely periodic then $A\in R_o$. We deduce the converse: Let $A\in R_o$. Then there are finitely many natural numbers n_i, $1\leq i\leq k$ and elements $A_i\in BG(n_i)$, $1\leq i\leq k$ such that A is a Boolean polynomial performed by the A_i, i.e. A is obtained by combining the A_i by unions and intersections. In this context observe that $A_i \in BG(n_i)$ iff its complement $-A_i := \mathbb{N} \setminus A_i$ is contained in $BG(n_i)$. Let n_o be the least common multiple of all n_i, $1 \le i \le k$. Then $A_i \in BG(n_o)$ for $1 \le i \le k$. This implies $A \in BG(n_o)$. Hence we conclude that $A \in R_o$ *iff A is purely periodic.* Consequently $R_o = \cup_{n \in \mathbb{N}} BG(n) = PER$.

We want to study the structure of R_o. For this reason we recall some notations concerning Boolean algebras: An indexed set $\{\mathfrak{A}_i\}_{i \in I}$ of subalgebras of the Boolean algebra $\mathbb{P}(\mathbb{N})$ is said to be *independent* provided $A_1 \cap \ldots \cap A_n \neq \emptyset$ for every finite sequence of non zero elements A_i chosen from subalgebras with pairwise different indices. Let $\mathfrak{B}$ be the Boolean algebra generated by the union of all $\mathfrak{A}_i$, $i \in I$. Then $\mathfrak{B}$ is called the *Boolean product* of the family $\{\mathfrak{A}_i\}_{i \in I}$ provided that this family is independent. The Stone space of $\mathfrak{B}$ is the topological product space of the Stone spaces of $\mathfrak{A}_i$, $i \in I$, i.e. $\text{Stone}(\mathfrak{B}) = \Pi_{i \in I} \text{Stone}(\mathfrak{A}_i)$ (cf. R.SIKORSKI [34] § 13 for proofs and details).

Some easy to describe Boolean algebras are introduced in the sequel. Their product will yield the proper Replica algebra R_o and a way to calculate the Stone space of R_o.

Let p be a prime number greater than one. If $k \in \mathbb{N}$ then we define

$$r_o(p^k, j) := \bigcup_{i=jp^{k-1}}^{(j+1)p^{k-1}-1} M[p^k, i](\mathbb{N})$$

where $j \in \mathbb{N} \cup \{0\}$, $0 \le j < p$. It follows that $\{r_o(p^k, j) \mid 0 \le j < p\}$ is a disjoint decomposition of $\mathbb{N}$. We denote $r_o(p^k)$ to be the Boolean subalgebra of $\mathbb{P}(\mathbb{N})$ which is generated by the class $\{r_o(p^k, j) \mid 0 \le j < p\}$. The Boolean algebra generated by all $r_o(p^k)$, $k \in \mathbb{N}$, is symbolized by $R_o(p)$.

For the sake of simplicity we explain the above notation by some examples: The even numbers are $r_o(2,0)$, and $r_o(2,1)$ are the odd numbers. By building 2-packages we obtain $r_o(2^2, 0) \triangleq (0,0;1,1;0,0;1,1;\ldots)$ and

$r_o(2^2,1)\cong(1,1;0,0;1,1;0,0;\ldots)$. Forming 2^2-packages, i.e. 4-packages, yields $r_o(2^3,0)$ etc. If $p=3$ then $r_o(3,0)\cong(0,0,1;0,0,1;0,0,1;\ldots)$, $r_o(3,1)\cong(0,1,0;0,1,0;0,1,0;\ldots)$, and $r_o(3,2)\cong(1,0,0;1,0,0;\ldots)$. By forming 3^1-packages we obtain $r_o(3^2,\ )$ etc. Observe that we represented a subset of $\mathbb{N}$ by its characteristic function, i.e. a 0-1 string.

It is obvious that the Boolean algebras $r_o(p^k)$ are independent. Hence $R_o(p)$ is their Boolean product: $R_o(p)=\Pi_{k=1}^{\infty} r_o(p^k)$. We mention the following useful property which is easy to verify: If $n\in\mathbb{N}$ then $\Pi_{k=1}^{n} r_o(p^k)$ is the Boolean subalgebra of $\mathbb{P}(\mathbb{N})$ which is generated by the following disjoint decomposition of $\mathbb{N}$: $\{M[p^n,i](\mathbb{N})\mid 0\leq i<p^n\}$. Therefore $\Pi_{k=1}^{n} r_o(p^k)=BG(p^n)$.

<u>PROPOSITION II.1.</u> Let $\{p_n\}_{n\in\mathbb{N}}$ be the indexed set of all primes greater than one. The proper Replica algebra R_o is the Boolean product of the indexed set $\{R_o(p_n)\}_{n\in\mathbb{N}}$ of Boolean algebras; i.e.

$$R_o=\Pi_{n=1}^{\infty} R_o(p_n)=\Pi_{n=1}^{\infty}\Pi_{k=1}^{\infty} r_o(p_n^k).$$

P r o o f: We prove that $\{R_o(p_n)\}_{n\in\mathbb{N}}$ is an independent set of Boolean subalgebras of $\mathbb{P}(\mathbb{N})$: Assume that $p_{n(1)},\ldots,p_{n(r)}$ are finitely many, pairwise different prime numbers. Let $\emptyset\neq A_{n(l)}\in R_o(p_{n(l)})$ for $1\leq l\leq r$. We have to prove that $A_{n(1)}\cap A_{n(2)}\cap\ldots\cap A_{n(r)}\neq\emptyset$. If $1\leq l\leq r$ then there is a $k(l)\in\mathbb{N}$ with $A_{n(l)}\in\Pi_{j=1}^{k(l)} r_o(p^j)$. Hence it suffices to prove that $M[p^{k(1)},i(1)](\mathbb{N})\cap\ldots\cap M[p^{k(r)},i(r)](\mathbb{N})\neq\emptyset$ where the system $\{i(l)\}_{l=1}^{r}$ is arbitrarily chosen with respect to $0\leq i(l)<p^{k(l)}$, $1\leq l\leq r$. Due to the theorem of simultaneous congruences (cf. G.H.HARDY and E.M.WRIGHT [12] theorem 121) the above inequality is always fulfilled. Hence $\{R_o(p_n)\}_{n\in\mathbb{N}}$ is an independent set of Boolean subalgebras of $\mathbb{P}(\mathbb{N})$.

We prove that $\Pi_{n=1}^{\infty} R_o(p_n)\subset R_o$: It suffices to prove that $R_o(p_n)\subset R_o$ for every $n\in\mathbb{N}$. If $n\in\mathbb{N}$ and $A\in R_o(p_n)$ then there is a k with $A\in\Pi_{j=1}^{k} r_o(p_n^j)=BG(p_n^k)$. Since $BG(p_n^k)\subset R_o$ the above inclusion is proven.

We prove that $R_o\subset\Pi_{n=1}^{\infty} R_o(p_n)$: Let $A\in R_o$. Then there is an $n\in\mathbb{N}$ with

$A \in BG(n)$. There are finitely many primes $p_{n(1)},\dots,p_{n(r)}$ and natural numbers $k(1),\dots,k(r)$ with $n=\Pi_{j=1}^{r} p_{n(j)}^{k(j)}$. It suffices to prove that $BG(n)=\Pi_{j=1}^{r} BG(p_{n(j)}^{k(j)})$. Remember that $BG(p_n^k)$ is generated by the class $\{M[p_n^k,i](\mathbb{N}) \mid 0 \leq i < p_n^k\}$. Therefore the above equality is obvious. This yields the proposition. Q.E.D.

The deduced product structure allows us to calculate the Stone space of R_o: Stone(R_o) is the topological product space of the family Stone $(r_o(p_n^k))$, $n,k \in \mathbb{N}$. It remains to calculate Stone$(r_o(p_n^k))$ for $n,k \in \mathbb{N}$.

The *Stone space* of a Boolean algebra $\mathfrak{A}$ is the unique compact space K with the following properties: (i) the clopen subsets of K are a neighbourhood basis system of K, (ii) the field CO(K) of clopen subsets of K is Boolean isomorphic to $\mathfrak{A}$.

This definitive characterization allows us to determine Stone$(r_o(p_n^k))$ for $n,k \in \mathbb{N}$: Let $\{0,\dots,p_n-1\}$ be the *discrete space* with p_n points, i.e. every point j, $0 \leq j \leq p_n-1$ is defined to be clopen. Then $\{0,\dots,p_n-1\}$ is the Stone space of $r_o(p_n^k)$ where $k \in \mathbb{N}$. A Boolean isomorphism $r_o(p_n^k) \xrightarrow{\simeq} CO(\{0,\dots,p_n-1\})$ is induced by $j \mapsto r_o(p_n^k,j)$ for $0 \leq j < p_n$. We denote $p_n^{\mathbb{N}}$ the countable topological product of $\{0,\dots,p_n-1\}$. Then we have proven the following result:

<u>PROPOSITION II.2.</u> Let $\{p_n\}_{n\in\mathbb{N}}$ be the indexed set of all primes greater than one. The Stone space of the proper Replica algebra R_o is

$$\text{Stone}(R_o)=\Pi_{n=1}^{\infty} p_n^{\mathbb{N}} = \Pi_{n=1}^{\infty} \Pi_1^{\infty} \{0,\dots,p_n-1\}.$$

We are prepared now to interpret the proper Replica algebra R_o by methods of measurement: The following questions define countably infinitely many methods of measurement: Is the agent $n \in \mathbb{N}$ contained in $r_o(2^k,0)$? Here $k \in \mathbb{N}$ fixes the method of measurement. There are countably infinitely many questions of the following kind: Is the agent $n \in \mathbb{N}$ contained in $r_o(3^k,0)$, $r_o(3^k,1)$ or $r_o(3^k,2)$? Again $k \in \mathbb{N}$ fixes the question. The possible outcome of such a question is $j \in \{0,1,2\}$ depending on $n \in r_o(3^k,j)$. Such a question defines a mapping from $\mathbb{N}$ into $\{0,1,2\}$. Its finite range coincides with our principle to measure by finite scales only. We conclude that such a three-valued question can be interpreted as a method of measurement. Remember that two-valued methods of measurement are used due to the technical simplicity only. If p_n is an arbitrary prime number then we obtain analogously p_n-valued methods of measurement. Hence the questions "Is the agent $m \in \mathbb{N}$ contained in $r_o(p_n^k, j)$?", where $k, n \in \mathbb{N}$, install $\mathbb{N} \times \mathbb{N}$ many methods of measurement. Due to G.Cantor $\mathbb{N} \times \mathbb{N}$ is bijective to $\mathbb{N}$. Therefore we obtain countably infinitely many methods of measurement.

A possible property combination with respect to the above questions yields a unique point in Stone(R_o) due to the product structure of Stone (R_o). The product topology on Stone(R_o) correponds to our principle to combine finitely many questions only. The realized property combinations are dense in Stone(R_o) due to the independentness of the questions. We conclude that the Boolean algebra R_o of all observable groups is in line with the principal grounding of the model developed in chapter one.

It may be strange that the measured groups $r_o(p_n^k, j)$ are mathematically constructed. We deem them to be justified for we have also interpreted the model as a completion of the class of all finite markets. Indeed in this chapter we combine the mathematical completeness requirement and the necessity that a microeconomic model has to be based on realistic, microeconomic measurement principles.

For the sake of conspicuousness we interpret Stone(R_o) explicitely as a compactification of $\mathbb{N}$ following our reasoning of chapter one.

We obtain a short explanation by remembering the original mathematical definition of the Stone space of the Boolean algebra R_o: A non empty subset $\mathfrak{F} \subset R_o$ is called a *filter* in R_o if $\emptyset \notin \mathfrak{F}$; $A \cap B \in \mathfrak{F}\ \forall A, B \in \mathfrak{F}$, and furthermore $A \in \mathfrak{F}$ for $B \in \mathfrak{F}$, $B \subset A$. A filter $\mathfrak{U} \subset R_o$ is an *ultrafilter* in R_o iff A or the complement $-A = \mathbb{N} \setminus A \in \mathfrak{U}$ for every $A \in R_o$. If $n_o \in \mathbb{N}$ then $\mathfrak{U}(n_o) := \{A \in R_o |$

$n_o \in A\}$ is the ultrafilter *determined by* n_o. We mention that there exist ultrafilters of a different type called free ultrafilters.

The set Stone(R_o) consists of all ultrafilters in R_o. Identifying $n \in \mathbb{N}$ with the ultrafilter $\mathcal{U}(n)$ determined by n we obtain $\mathbb{N}$ as a subset of Stone(R_o).

Let $A \in R_o$. Then the set $\bar{A} := \{\mathcal{U} \mid A \in \mathcal{U}$ and $\mathcal{U}$ is an ultrafilter in $R_o\} \subset$ Stone(R_o) is defined to be closed and open i.e. *clopen*. Due to this definition one can deduce that Stone(R_o) is a compact space having $\{\bar{A} \mid A \in R_o\}$ as a neighbourhood basis system. It turns out that this neighbourhood basis system consists exactly of all clopen subsets of Stone(R_o) (see R.SIKORSKI [35] for detailed proofs). The obtained compact space Stone (R_o) is called the *Stone space* of R_o.

Due to this abstract approach one recognizes that $A \mapsto \bar{A}$ installs the *Boolean isomorphism* between R_o and the system CO(Stone(R_o)) of clopen subsets of Stone(R_o). Since $n \in \mathbb{N}$ was identified with $\mathcal{U}(n)$ it follows that $\mathbb{N}$ is a *dense subset* of Stone(R_o). Hence Stone(R_o) *is a compactification of* $\mathbb{N}$. With respect to the above topology one recognizes that $\bar{A}$ is the *closure* cl(A) of $A \in R_o$.

3.3 THE UNIQUE NATURAL INFLUENCE MEASURE. Due to the interpretation of the idealized, concrete model as a completion we could specify the unique Boolean algebra R_o of all observable groups. The same principle will install a unique natural influence measure on R_o.

A finite market was interpreted by an infinite replication as an idealized market. This implies that the natural influence measure on every BG(n), $n \in \mathbb{N}$ is necessarily the density d. We deduced that $R_o = \bigcup_{n \in \mathbb{N}} BG(n)$. Consequently *the density* d *is the unique natural influence measure* defined on R_o, i.e. if $A \in R_o$ then

$$d(A) = \lim_{n \to \infty} \frac{1}{n} \sum_{i=1}^{n} \chi_A(i)$$

where χ_A denotes the characteristic function of A. It remains to verify that d is a normalized, non negative, finitely additive measure on R_o. But this is obvious for $R_o = \cup_{n\in\mathbb{N}} BG(n)$.

We prove that *pure competition holds with respect to* R_o *and* d: Let $\varepsilon > 0$ and let $n \in \mathbb{N}$ with $1/n < \varepsilon$. Then $\{M[n,i](\mathbb{N}) \mid i \in \mathbb{N} \cup \{0\},\ i<n\}$ is a finite, disjoint decomposition of $\mathbb{N}$ by elements of R_o and $d(M[n,i](\mathbb{N})) = 1/n < \varepsilon$ for $0 \leq i < n$. Therefore pure competition holds.

In the next step we want to calculate the probability measure $\bar{d}$ induced by the density d on Stone(R_o). We will recognize that $\bar{d}$ is just the canonical product measure on Stone(R_o).

Let p be a prime number greater than one. Let $k \in \mathbb{N}$, and let $j \in \mathbb{N} \cup \{0\}$ with $0 \leq j < p$. Due to the definition of the density and of $r_o(p^k,j)$ we obtain $d(r_o(p^k,j)) = 1/p$. The Stone space of the Boolean algebra $r_o(p^k)$ was the discrete space $\{0,1,\ldots,p-1\}$. The mapping $j \mapsto r_o(p^k,j)$ has induced a Boolean isomorphism between $r_o(p^k)$ and the clopen subsets - i.e. here *all* subsets - of Stone($r_o(p^k)$). Therefore we recognize that d induces on Stone($r_o(p^k)$) the Laplace probability l_p i.e. the canonical, equally distributed probability measure.

We prove that the density d induces on $\mathrm{Stone}(R_o(p)) = \Pi_1^\infty \{0,1,\ldots,p-1\}$ the canonical product measure derived from the Laplace probability l_p on $\{0,1,\ldots,p-1\}$: Keep in mind that the Boolean isomorphism $R_o(p) \cong CO(\mathrm{Stone}(R_o(p)))$ is obtained by the product of the canonical Boolean isomorphisms $r_o(p^k) \cong \{0,1,\ldots,p-1\}$ (cf. R.SIKORSKI [35] § 13). Let $k(1),\ldots,k(m)$ be pairwise different natural numbers. Let $j(1),\ldots,j(m)$ be integers with $0 \leq j(1),\ldots,j(m) < p$. Due to the definition of $r_o(p^k,j)$ one easily verifies the following equation: $d(\cap_{l=1}^m r_o(p^{k(l)},j(l))) = d(r_o(p^{k(1)},j(1))) \cdot \ldots \cdot d(r_o(p^{k(m)},j(m)))$.[2)] Therefore the density d induces on CO(Stone($R_o(p)$)) the

---

2) This means that the Boolean algebras $r_o(p^k), k \in \mathbb{N}$ are not only independent but even d-independent (see D.A.VLADIMIROV [36] for this canonical approach).

canonical product measure. Due to the usual extension procedure we obtain on $\Pi_1^\infty\{0,1,\ldots,p-1\}$ the canonical product probability measure derived from the Laplace probability 1_p.

We prove that d induces on $\text{Stone}(R_o)=\Pi_{n=1}^\infty\text{Stone}(R_o(p_n))$ the product measure derived from the induced product measures on $\text{Stone}(R_o(p_n)),n\in\mathbb{N}$. Here $\{p_n\}_{n=1}^\infty$ denotes again the sequence of all primes greater than one. In analogy to the previous step it suffices to prove the following: Let $n(1),\ldots,n(m)$ be pairwise different natural numbers. Let $A_i\in R_o(p(i))$ for $1\leq i\leq m$. Then $d(\cap_{i=1}^m A_i)=d(A_1)\cdot\ldots\cdot d(A_m)$ has to hold. For a verification of that equality observe the following: For $1\leq i\leq m$ there is a natural number $k(i)$ such that A_i possesses the period $p(i)^{k(i)}$. This implies that A_i is the finite disjoint union of sets of the following kind: $M[p(i)^{k(i)},j(i)](\mathbb{N})=\{p(i)^{k(i)}r-j(i)\mid r\in\mathbb{N}\}$ where $0\leq j(i)<p(i)^{k(i)}$. Therefore it is sufficient to verify the above equation for these special sets, i.e. we have to prove that $d(\cap_{i=1}^m M[p(i)^{k(i)},j(i)](\mathbb{N}))=d(M[p(1)^{k(1)},j(1)](\mathbb{N}))\cdot\ldots\cdot d(M[p(m)^{k(m)},j(m)](\mathbb{N}))$ holds for $k(i)\in\mathbb{N}$ and $0\leq j(i)<p(i)^{k(i)}$. But this last equation is a consequence of the theorem on simultaneous congruences (cf. G.H.HARDY and E.M.WRIGHT [12] theorem 121). Therefore we obtain the following result:

PROPOSITION II.3. Let $\{p_n\}_{n\in\mathbb{N}}$ be the indexed set of all primes greater than one. The density d is a normalized, non negative finitely additive measure defined on the proper Replica algebra R_o. The Radon probability measure $\bar{d}$, induced by d on $\text{Stone}(R_o)=\Pi_{n=1}^\infty\Pi_1^\infty\{0,\ldots,p_n-1\}$, is the canonical product measure deduced from the Laplace probabilities 1_{p_n} on the discrete spaces $\{0,1,\ldots,p_n-1\}$. The Radon measure $\bar{d}$ is atomless.

Let K be a compact space endowed with a non negative Radon measure μ. Then $\text{supp}(\mu):=\{x\in K\mid\mu(U)>0$ for every open neighbourhood U of $x\}$ denotes the *support* of μ.

REMARK II.4.a. The support $\text{supp}(\bar{d})$ of the Radon measure $\bar{d}$ is the whole space $\text{Stone}(R_o)$.

It remains to prove that $\mathbb{N}$, interpreted as a subsequence of Stone (R_o), is $\bar{d}$-*uniformly distributed*. This requires us to prove the following: If $\bar{f}$ is a real valued, continuous function on Stone (R_o) - where f denotes the restriction of $\bar{f}$ onto $\mathbb{N}$ - then

$$(+) \qquad \lim_{n\to\infty} \frac{1}{n} \sum_{i=1}^{n} f(i) = \int_{\text{Stone}(R_o)} \bar{f} d\bar{d}.$$

Let $\bar{A} \subset \text{Stone}(R_o)$ be clopen. Then $A := \bar{A} \cap \mathbb{N} \in R_o$ and $\text{cl}(A) = \bar{A}$. Moreover $d(A) = \bar{d}(\bar{A})$. This implies that (+) holds for the characteristic function $\chi(\bar{A})$ of $\bar{A}$. Let $\bar{f}$ be a real valued, continuous function on Stone(R_o), and let $\varepsilon > 0$. Then $\bar{f}$ is uniformly continuous on Stone(R_o). Since the field of clopen sets is a neighbourhood basis system of Stone(R_o) there is a disjoint decomposition $\{\bar{A}_i\}_{i=1}^{m}$ of Stone(R_o) by clopen sets and real numbers a_i; $i=1,\dots,m$ such that $\|\sum_{i=1}^{m} a_i \chi(\bar{A}_i) - \bar{f}\| = \sup\{|\sum_{i=1}^{m} a_i \chi(\bar{A}_i)(x) - \bar{f}(x)| \mid x \in \text{Stone}(R_o)\} < \varepsilon$. Since (+) holds for $\sum_{i=1}^{m} a_i \chi(\bar{A}_i)$ and since $\varepsilon > 0$ was arbitrarily chosen, the formula (+) holds for $\bar{f}$. Hence we obtain the

REMARK II.4.b. The set $\mathbb{N}$ of agents, interpreted as a subsequence of Stone(R_o), is $\bar{d}$-uniformly distributed.

We can state here that the requirement to interpret *every* finite market as a special case of an idealized market has solved two open questions of the first chapter in a natural way: i.e. we have obtained the concrete Boolean algebra R_o of observable groups and we have obtained the unique, precisely determined natural influence measure d on R_o. Both are characterized by a well-to-calculate product structure.

3.4 DETERMINISTIC MARKETS. We start by discussing the meaning of an endowment mapping in the deterministic concrete model. Later on market mappings and markets will be concretized.

Due to our reasoning in the general part it seems to be natural that every continuous mapping $\bar{\mathcal{E}} : \text{Stone}(R_o) \to P \times \mathbb{R}^n_+$ is accepted in the deterministic concrete model as an extended endowment mapping. Although we will finally proceed so it is necessary to scrutinize the definition with respect to the here underlying completion concept.

The class of all admitted market mappings shall be a completion of the class of all finite markets where a finite market here is interpreted as its idealized, replicated counterpart defined on $\mathbb{N}$. Therefore we describe at first which mappings $\bar{\mathcal{E}}:\text{Stone}(R_o)\longrightarrow P\times\mathbb{R}^n_+$ correspond to replicated endowment mappings $\mathcal{E}:\mathbb{N}\longrightarrow P\times\mathbb{R}^n_+$:

Assume that $\mathcal{E}:\mathbb{N}\longrightarrow P\times\mathbb{R}^n_+$ is a replicated endowment mapping. Then there is a $n\in\mathbb{N}$ and a disjoint decomposition $\{A_i\}_{i=1}^m\subset BG(n)$ of $\mathbb{N}$ such that $\mathcal{E}$ is constant on every A_i. Let $\bar{A}_i\subset\text{Stone}(R_o)$ be the clopen set corresponding to A_i via the canonical Boolean isomorphism $R_o\overset{\sim}{\Longrightarrow}CO(\text{Stone}(R_o))$. Then $\{\bar{A}_i\}_{i=1}^m$ is a disjoint decomposition of $\text{Stone}(R_o)$ by clopen sets. Then $\mathcal{E}$ yields the mapping $\bar{\mathcal{E}}:\text{Stone}(R_o)\longrightarrow P\times\mathbb{R}^n_+$ as follows: $\bar{\mathcal{E}}\big|_{\bar{A}_i}:=\mathcal{E}\big|_{A_i}$, $1\leq i\leq m$. Remember that a mapping $\mathcal{F}$ having a finite range is called *simple*. We recognize that the replicated endowment mapping $\mathcal{E}$ yields the continuous, simple mapping $\bar{\mathcal{E}}$ defined on $\text{Stone}(R_o)$.

Let $\bar{\mathcal{E}}:\text{Stone}(R_o)\longrightarrow P\times\mathbb{R}^n_+$ be a continuous simple mapping. Then there is a disjoint decomposition $\{\bar{A}_i\}_{i=1}^m$ of $\text{Stone}(R_o)$ by clopen sets such that $\bar{\mathcal{E}}$ is constant on every $\bar{A}_i$. The canonical Boolean isomorphism yields the disjoint decomposition $\{A_i\}_{i=1}^m\subset R_o$ of $\mathbb{N}$. If n is the least common multiple of the periods of A_i, $1\leq i\leq m$ then $\{A_i\}_{i=1}^m\subset BG(n)$. We define $\mathcal{E}:\mathbb{N}\longrightarrow P\times\mathbb{R}^n_+$ by $\mathcal{E}\big|_{A_i}=\bar{\mathcal{E}}\big|_{\bar{A}_i}$; where $1\leq i\leq m$. Then $\mathcal{E}$ is a replicated endowment mapping.

Hence there is a natural *bijection between the replicated endowment mappings and the* $P\times\mathbb{R}^n_+$*-valued, simple, continuous mappings* defined on $\text{Stone}(R_o)$. One easily verifies that this natural bijection is *isometric* with respect to the sup.metric.

We want to interpret the above Boolean deduced isometric bijection by topological arguments: Remember that $\mathbb{N}$ was considered as a dense subset of Stone(R_o). Within this framework we deduced that the closure cl(A) of $A \subset \mathbb{N}$, $A \in R_o$ corresponds to the image $\bar{A}$ of A with respect to the canonical Boolean isomorphism between R_o and CO(Stone(R_o)); i.e. we obtained $\bar{A}$=cl(A) for $A \in R_o$. Then it is obvious that the above stated isometric bijection $\mathcal{E} \to \bar{\mathcal{E}}$ is obtained by the *continuous extension* of $\mathcal{E}$ - defined on $\mathbb{N} \subset$ Stone(R_o) - to $\bar{\mathcal{E}}$, defined on Stone(R_o).

PROPOSITION II.5. Let K be a compact space having a neighbourhood basis system of clopen sets. Let (M,d) be a metric space with the metric d. The space C(K,M) of continuous, M-valued functions defined on K is the completion of the continuous, simple, M-valued functions on K with respect to the sup metric.

P r o o f: Remember that the space C(K,M) is complete with respect to the sup.metric (cf. K.KURATOWSKI [25] § 21 and § 33). The class of continuous, simple, M-valued functions defined on K is contained in C(K,M). Hence it suffices to prove that every $f \in C(K,M)$ can be approximated by simple functions with respect to the sup metric. Since K is zero dimensional - i.e. K possesses a neighbourhood basis system consisting of clopen sets - and since f is uniformly continuous this is obviously possible. Q.E.D.

The space $P \times \mathbb{R}^n_+$ was endowed with the metric $\delta \times \|\ \|$ where δ was the closed convergence metric on P and where $\|\ \|$ was the Euclidean metric on $\mathbb{R}^n_+$. On C(Stone(R_o),$P \times \mathbb{R}^n_+$) we considered the sup metric d_o defined by $d_o(\bar{\mathcal{E}},\bar{\mathcal{F}}) = d_o(\bar{E} \times \bar{e}, \bar{F} \times \bar{f}) := \sup\{\delta(\bar{E}(x),\bar{F}(x)) + \|\bar{e}(x) - \bar{f}(x)\| \mid x \in \text{Stone}(R_o)\}$ where $\bar{\mathcal{E}}, \bar{\mathcal{F}} \in$ C(Stone(R_o),$P \times \mathbb{R}^n_+$). Due to prop. II.5 we obtain C(Stone(R_o),$P \times \mathbb{R}^n_+$) as the completion of the extended, replicated endowment mappings with respect to the sup metric d_o.

A continuous mapping defined on a compact space is uniformly continuous. Uniformly continuous mappings on a compact space can be treated equivalently on any dense subset of that compact space via the restriction - resp. extension - procedure (for this standard topological argument cf. e.g. K.KURATOWSKI [25]).

Due to the foregoing conclusions we recognize that $C(\mathrm{Stone}(R_o), P\times\mathbb{R}^n_+)$ is a completion of the extended, replicated endowment mappings which can be treated equivalently on the set $\mathbb{N}$ of agents. Therefore $C(\mathrm{Stone}(R_o), P\times\mathbb{R}^n_+)$ seems to be the appropriate class of admitted endowment mappings for our model. However, since a completion is directly dependent of the underlying metric, it remains to justify the use of the sup.metric d_o for the completion process.

Remember the following, principal construction ideas of the considered, deterministic replica model: The model is understood as a mathematical completion of the class of finite markets. The completion procedure has to respect the general, logical foundation developed in chapter one. Consequently, the functions defined on $\mathrm{Stone}(R_o)$ are seen as a result of a *deterministic* observation procedure. In a deterministic observation *every* agent has his own importance. Therefore we conclude that the use of the sup metric is well justified for no agent is neglected by a deterministic observation.

The reader may argue that the sup metric yields a contradiction to the idea of pure competition, where no single agent possesses any influence on the market. But observe that until this moment we developed a deterministic observation procedure and - in certain separate sense - the meaning of an influence measure. We have never applied the influence measure onto the observation process. Hence the above argument does not challenge the use of the sup metric. It states simply that until this moment the situation of pure competition was never directly treated. Indeed this is reserved for the next parts.

We are able now to concretize the definition of a market from chapter one:

<u>DEFINITION II.6.</u> Let R_o be the proper Replica algebra and let d be the density defined on R_o. (i) A continuous mapping $\bar{\mathcal{E}} = \bar{E}\times\bar{e}: \text{Stone}(R_o) \to P\times\mathbb{R}^n_+$ is called a deterministic, observed endowment mapping. (ii) Let $\bar{E}(\text{Stone}(R_o)) \subset P_{mo}$. Then $\bar{\mathcal{E}}$ is a deterministic, observed market mapping. The restriction $\mathcal{E}$ of $\bar{\mathcal{E}}$ onto $\mathbb{N}$ is a deterministic endowment mapping, resp. a deterministic market mapping. (iii) A deterministic, observed market with pure competition is symbolized by $\bar{\mathcal{E}}: (\text{Stone}(R_o), CO(\text{Stone}(R_o)), \bar{d}) \to P_{mo}\times\mathbb{R}^n_+$ where $\bar{\mathcal{E}}$ is analogous as in (ii). By the restriction onto $\mathbb{N}$ we obtain a deterministic market $\mathcal{E}: (\mathbb{N}, R_o, d) \to P_{mo}\times\mathbb{R}^n_+$ with pure competition.

Observe that a deterministic market mapping fulfills the stability assumptions from chapter I 2.2 due to its definition by a restriction.

We still have the duty to inform the reader about a small gap between the scanning picture, developed in chapter one, and the concept defined here of a deterministic market:

In the first chapter we were dealing with the observation of *one single* market mapping. Then, by a scaling, we derived a system of coalitions with respect to that special market mapping. Hence the coalitions were dependent on the considered market mapping in the following sense: Every observed market mapping corresponded to a simple mapping on the generated Stone space.

The procedure in this section was somewhat different. We started with the class of all replicated markets. Then we developed the system R_o of coalitions such that all replicated markets could be interpreted as simple mappings on the Stone space of R_o. Hence *every* replicated market mapping could be treated analogously as in the first chapter with respect to the system of coalitions R_o.

Due to the completeness concept we admitted in a further step continuous mappings on Stone(R_o) as observed market mappings. Hence *we added a large class of market mappings* merely following a mathematical requirement. If we would scan an arbitrary continuous mapping in the same manner as in the first chapter, then we would generate in general coalitions *not contained* in the proper Replica algebra R_o.

Concerning the scanning picture the above behaviour of the deterministic replica model is an inconsistency which we have to clarify: The scanning tool was developed to explain the observation of *one single* market.[3] The completion tool was introduced to *select a class* of market mappings which is sufficient for our purposes and which possesses suitable completeness properties. Therefore it is less surprising that the resulting class of admitted market mappings is not always interpretable in a direct way. Indeed this behaviour is canonically inherent to the powerful, but *abstract, completion approach* as the following analogue stresses:

The rational numbers are easy to interpret from the viewpoint of reality. The rational numbers are dense in the complete metric space of real numbers. A real number can be seen as a converging sequence of rational numbers. However a real number itself is in general not directly interpretable within reality. With respect to our framework the R_o-simple mappings are easy to interpret, and they are dense in the complete space of all continuous endowment mappings. In particular, a continuous endowment mapping can be seen as a converging sequence of simple mappings. An arbitrary continuous endowment mapping is in general not directly interpretable as an R_o-scanning.

Due to the above direct analogue we deem it to be justified that continuous functions on Stone(R_o) are admitted in the sequel as deterministic market mappings.

---

[3] The concept of chapter one can be applied analogously with respect to a class of countably many markets. If uncountably many markets are observed simultaneously, then we obtain in general an uncountable system of generated coalitions. The class of continuous functions on Stone(R_o) is not countable.

In this stage of the model we can present the argument which stimulated us to restrict the analysis in chapter one onto *bounded* functions: We have interpreted the model as an abstract observation. Simultaneously we consider it now as a completion. In our foregoing reasoning we deduced that the continuous mappings are the appropriate class of idealized markets. Since the continuous functions on a compact space are especially bounded functions, we conclude that the *completion principle requires the restriction onto bounded mappings*.

3.5 CONCLUSION. In chapter one we transformed the deterministic model of a large market with countable many agents into a continuous, deterministic model. This continuous counterpart was justified, motivated, and explained by a scanning procedure. Its familiarity with related measurements in reality was demonstrated. It was our goal to use the interaction between reality and abstraction as far as possible to install a microeconomic consistent interpretation of the necessarily abstract concept of pure competition.

The status of pure competition is an abstract, economic one. It is an abstraction just like the status of a vacuum is in an idealized concept in physics. In analogy to the status of a vacuum, the status of pure competition can never be reached in reality. It may be approximated up to a certain degree in special situations of reality. Therefore there is an intrinsic *economic* requirement to interpret a market with pure competition as a limit of finite markets having an approximated behaviour of pure competition. The limit-interpretation, due to F.Y.EDGEWORTH [10], yields a powerful interaction between reality and the abstract modelling. This interaction is a second one different to that which was elaborated in the first chapter.

Due to the abstract starting level for our analysis in chapter one - we considered countably many agents - there remained necessarily a lack of concreteness concerning the system of coalitions and concerning the natural influence measure. The requirement, to interpret the resulting abstract model as a limit of finite markets, yielded us the powerful *mathematical completion* tool as the appropriate means to fill the remaining gap:

The system of all purely periodic subsets of $\mathbb{N}$ was installed as the class R_o of all coalitions. The limit of the Laplace probabilities turned out to be well defined and atomless with respect to the Boolean algebra R_o. Consequently this limit was declared to be our unique natural influence measure. Every $P \times \mathbb{R}^n_+$ valued, replicated mapping was admitted as an endowment mapping. With respect to the scanning picture, an interpretation of every such endowment mapping was possible with the help of the fixed Boolean algebra R_o. Following the completeness goal we admitted every $P \times \mathbb{R}^n_+$ — valued, continuous function on Stone(R_o) as an admitted, extended endowment mapping. The endowment mappings obtained in this way could not be interpreted in general by the scanning picture with respect to the fixed Boolean algebra R_o. However every continuous endowment mapping was shown to be the limit of some replicated endowment mappings. Therefore we admitted the continuous mappings in the same manner as we accepted e.g. the real numbers and not only the rationals. Due to the limit interpretation of our model - i.e. due to our restriction onto continuous mappings - we could justify our restriction onto bounded functions in chapter one.

Therefore, by adding the monotonicity definition from chapter one, we obtained a *carefully justified, concrete definition of a deterministic market with pure competition.*

The model, which we have developed so far, is still a deterministic one. Indeed it consists of two, rather separated parts. The one part is the unique natural influence measure and the method to calculate it. The other is given by the admitted market mappings and the deterministic 'observation-completion' approach by which they are explained and justified.

We reported that every public opinion poll is based on a deterministic method of measurement. Therefore the use of a system of methods of measurement implies a *deterministic analysis* of their interaction and combination. As long as this analysis is not finished the separation of the above mentioned major parts of the entire model can not be avoided.

The problem of pure competition as well as the observation by public opinion polls require probabilistic considerations. Therefore the reader may expect that the model will change strongly in the sequel. We mention here that this will not be the case. Indeed an astonishingly slight reinterpretation of the elaborated results will connect the two, separated parts of the model in a natural, probabilistic manner. The installed, deterministic framework will remain the *stable and concrete grounding* for the following probabilistic model.

4. THE PROBABILISTIC REPLICA MODEL

In the course of the first three paragraphs we accomplished the deterministic analysis of a large market. Following our intuition, borrowed from the theory of public opinion polls, we will subjoin elements of probability theory to receive the final definition of a market with pure competition. A mathematical isomorphism result will enable us to present this definition in an elementary manner.

4.1 MARKETS WITH PURE COMPETITION. In the first chapter we observed markets in a deterministic way by methods of measurement. In the sequel we want to observe markets in a *probabilistic* way by public opinion polls.

If we inspect a market by a method of measurement φ_G, then we obtain as our result the group $G_\varphi:\{n\in N \mid \varphi_G(n)=1\}$; i.e. we get the well determined coalition $G_\varphi \subset N$. Setting aside the idealized simplification of our two-valued methods of measurement - which was introduced due to technical reasons - every public opinion poll is *based* on a method of measurement. Therefore the principal reasonning of the first chapter is *not* necessarily abolished. We have to take into account in the following the less precise information which a public opinion poll delivers us.

The numerical value of $\nu(G_\varphi)$ remains unaltered if G_φ is varied by a subset $A\subset N$ with $\nu(A)=0$. Therefore a question φ together with the numerical value $\nu(G_\varphi)$ yields no knowledge of the group G_φ. We can at best conclude on the related *equivalence class of groups*, i.e. on $\{G\subset N \mid \nu(G\Delta G_\varphi)=0\}$. Therefore we have to regard in the sequel that the topics of our analysis are determined up to zero influence sets only. This implies a study of the class of all ν-zero subsets of N. We present a further, economic reason for such an investigation:

Due to the assumption of pure competition the influence of a single agent onto the economic situation is zero in our model. Therefore an alteration of a market mapping on a zero influence set of agents - e.g. one single agent - cannot yield any alteration of the economic situation. Therefore a market mapping is in a natural way an *equivalence class of mappings*. Once more this equivalence class is determined by the class of all ν-zero sets resp. by the class of all subsets $\mathbb{A}$ of $\mathbb{N}$ with $d(\mathbb{A})=0$. Hence we have to scrutinize the class of zero influence sets in the sequel.

The construction of the Boolean algebra of coalitions R_o, resp. $BG(\Phi)$, was not motivated by any interest in zero influence sets. Therefore we cannot await that this class is completely contained in $BG(\Phi)$ resp. R_o. Indeed d defined on R_o is *strictly positive*, i.e. if $A\neq\emptyset$,

$A \in R_o$ then $d(A)>0$. Consequently it will be necessary to change to enlarged Boolean algebras of coalitions. We shall expect that measure theoretic tools are helpful to solve our problems.

Without concretizing the class of zero influence subsets of N we claim some of its properties: Provided that pure competition holds, every single agent is a zero influence set. The whole set N, resp. $\mathbb{N}$, of agents is no set of zero influence. We conclude that the class of zero influence sets cannot form a Boolean σ-ideal for N is countable. Therefore *the σ-additive Lebesgue measure theory is excluded* to put the probabilistic character into the model.

Insteed of the σ-additive measure theory the *generalized Jordan content* and *the Riemann-Darboux integration theory* will be applied in our model. This theory was developed by S.ROLEWICZ and the author in [15], [16], [17], and [18]. A comprehensive introduction is contained in [18] resp. in the appendix. We will discuss several special properties of that theory to demonstrate its aptitude with respect to the problems arising here.

Let K be a compact space endowed with a normalized, non negative Radon measure μ. A subset $A \subset K$ is a μ-*continuity set* iff its boundary $\partial A = cl(A) \cap cl(K \setminus A)$ has μ-measure zero. The μ-*content* μ^c is the restriction of the completion of μ onto *the field* $\mathcal{B}(K,\mu)$ *of all* μ-*continuity sets*; e.g. if K is the unit interval [0,1] and if μ is the Lebesgue measure λ then $\mathcal{B}([0,1],\lambda)$ are the Jordan measurable sets and λ^c is the Jordan content. The following property of the generalized Jordan content, consistent with our previous modelling, has to be mentioned: *A subset* $A \subset K$ *is a* μ^c*-zero set (also called: a* μ*-content zero set or a* μ*-Jordan zero set) iff* $\mu(cl(A))=0$.

The rational subintervals of [0,1] suffice to calculate the classical Riemann integral. The subsequent notation will install an analogue for the general case: Let $\mathbb{B}(K,\mu) \subset \mathcal{B}(K,\mu)$ be a field fulfilling: (i)

$\mathbb{B}(K,\mu)$ contains a neighbourhood basis system of K and (ii) $A\in\mathbb{B}(K,\mu)$ implies $cl(A)\in\mathbb{B}(K,\mu)$. Then $\mathbb{B}(K,\mu)$ is called an *integral* μ-*basis field*; e.g. the clopen subsets of a Stone space are an integral basis field with respect to every non negative Radon measure on that Stone space. If $K=Stone(R_o)$ then *the field* $\overline{\mathcal{R}}_o$, *generated by the clopen sets and the* $\bar{d}^c$-*zero sets*, is an integral $\bar{d}^c$ basis field. In particular $\overline{\mathcal{R}}_o$ *is adequate to consider equivalence classes of groups* as it was required at the beginning of this section.

Let S be a dense subset of K. We denote $\mathbb{B}(K,\mu,S)=\mathbb{B}(K,\mu)\cap S$. We define the (induced, topological) μ-$\mathbb{B}$-*trace content* $\mu_S^{\mathbb{B}}$ on $\mathbb{B}(K,\mu,S)$ by $\mu_S^{\mathbb{B}}(A):=\mu(cl(A))$. Then $\mu_S^{\mathbb{B}}$ is a normalized, non negative, finitely additive measure defined on the field $\mathbb{B}(K,\mu,S)$. Then $\mathbb{B}(K,\mu,S)$ divided by the $\mu_S^{\mathbb{B}}$-zero sets is *measure preserving, Boolean isomorphic to* $\mathbb{B}(K,\mu)$ divided by the $\mu_S^{\mathbb{B}}$-zero sets. The *intersection* mapping - resp. the *closure* mapping - induces this isomorphism. Here $\mu^{\mathbb{B}}$ denotes the restriction of μ^c onto $\mathbb{B}(K,\mu)$. We conclude that *the generalized Jordan content can be treated equivalently on dense subsets of* K. This fact is exactly in line with requirement (A5).

<u>EXAMPLE II.1.</u> Let $K=Stone(R_o)$, let $\mu=\bar{d}$, and let $S=\mathbb{N}$. Then $\bar{d}^c_{\mathbb{N}}$ is the density d defined on a field containing R_o. If we choose $\mathbb{B}(Stone(R_o),\bar{d})=\overline{\mathcal{R}}_o$ then $\bar{d}_{\mathbb{N}}^{\mathbb{B}}$ is the density d defined on the field which is generated by R_o and the subsets $A\subset\mathbb{N}$ with $\bar{d}(cl(A))=0$. *We denote this field* $\mathcal{R}_o$ in the sequel. Obviously $\mathcal{R}_o=\overline{\mathcal{R}}_o\cap\mathbb{N}$. The density d was defined on R_o by the limit of finite average means. Due to a general result we will see soon that this formula holds true as well for $\mathcal{R}_o$ (see prop. II.7 (iii)).

We demonstrated by several properties that the generalized Jordan content is appropriate for our purposes. In the sequel we deduce in the same concentrated manner that the correlated generalization of B.Riemann's integration theory is analogously well adapted to our modelling.

As an integration calculus is meaningful only on the support of a measure, we *assume* in the following that $\mathrm{supp}(\mu)=K$. Remember that $\mathrm{supp}(\bar{d})=\mathrm{Stone}(R_o)$.

A $\mathbb{B}(K,\mu)$-*partition* of K is a finite class $\mathcal{P}=\{P_i\}_{i=1}^n\subset\mathbb{B}(K,\mu)$ consisting of non void, regular closed sets with $\cup_{i=1}^n P_i=K$ and $\mu(P_i\cap P_j)=0$ for $i\neq j$. A subset $A\subset K$ is *regular closed* iff $A=\mathrm{cl}\,\mathrm{int}(A)$. Analogously as in the classical situation the $\mathbb{B}(K,\mu)$-partitions are *directed* via the inclusion. We denote $(S\mathbb{B}\mathcal{P},\geq)$ the directed class of all $\mathbb{B}(K,\mu)$-partitions. Due to the intersection resp. closure mapping $\mathbb{B}(K,\mu)$-partitions can be considered equivalently with respect to $\mathbb{B}(K,\mu,S)$. This behaviour is consistent with requirement (A5).

We consider $\mathbb{R}^n$ endowed with a norm $\|\ \|$ e.g. the Euclidean norm. Let $f:K\longrightarrow\mathbb{R}^n$ be a bounded function. If $M\subset\mathbb{R}^n$ then $\mathrm{dia}(M):=\sup\{\|x-y\|\ |x,y\in M\}$ denotes the *diameter* of M. Let $\mathcal{P}=\{P_i\}_{i=1}^n$ be a $\mathbb{B}(K,\mu)$-partition of K. Then $\mathrm{dis}(f,\mathcal{P}):=\sum_{i=1}^n \mathrm{dia}(f(P_i))\mu^c(P_i)$ denotes the μ-*distance sum* of f with respect to $\mathcal{P}$. We call f μ-*Darboux integrable* iff $\inf\{\mathrm{dis}(f,\mathcal{P})\ |\ \mathcal{P}\in S\mathbb{B}\mathcal{P}\}=0$. All definitions can be applied analogously if $\mathbb{R}^n$ is replaced by a compact metric space e.g. P with the closed convergence metric δ.

If $x_i\in f(P_i)$ for $i=1,\dots,n$ then $x=\sum_{i=1}^n\mu^c(P_i)x_i$ is called a $\mathcal{P}$-*sum* of f. We denote $S(f,\mathcal{P}):=\{x\in\mathbb{R}^n\ |x \text{ is a } \mathcal{P}\text{-sum of } f\}$. If the indexed class $\{S(f,\mathcal{P})\ |\ \mathcal{P}\in(S\mathbb{B}\mathcal{P},\geq)\}$ is converging then

$$^R\!\!\int_K f\,d\mu:=\lim_{\mathcal{P}\in(S\mathbb{B}\mathcal{P},\geq)} S(f,\mathcal{P})$$

is called the μ-*Riemann integral* of f over K.

The above definitions are generalizations of the classical ones. The class of μ-Darboux integrable functions and the μ-Riemann integral are independent of the chosen integral μ-basis field. They can be considered equivalently on a dense subset of K. A bounded function f is μ-Darboux integrable iff its μ-Riemann integral exists. We mention the following analogue of the classical result (for a proof see [18] and

[15] of S.ROLEWICZ and the author):

PROPOSITION II.7. Let K be the compact support of the normalized, non negative Radon measure μ. Let $f:K \longrightarrow \mathbb{R}^n$ be bounded. (i) f is μ-Darboux integrable iff f is continuous μ-almost everywhere. (ii) If f is μ-Darboux integrable then f is $\bar{\mu}$-Lebesgue integrable and the μ-Riemann integral of f coincides with its $\bar{\mu}$-Lebesgue integral (here $\bar{\mu}$ denotes the completion of μ). (iii) If $\{x_i\}_{i\in\mathbb{N}}$ is μ-uniformly distributed in K then

$$\lim_{n\to\infty} \frac{1}{n} \sum_{i=1}^{n} f(x_i) = {}^{R}\!\!\int_K f d\mu$$

provided that f is μ-Darboux integrable.

The prop. II.7 (i) holds analogously if $\mathbb{R}^n$ is replaced by a compact metric space. Since prop. II.7 (i) is a finitely additive result, formulated in the σ-additive language, it cannot be applied directly with respect to a dense subset S of K. A different, equivalent characterization allows that treatment. We noted prop. II.7 (i) mainly to emphasis the similarity between the classical and the generalized theory.

Due to our introductory explanations we are mainly interested in a calculus of equivalence classes suitable for our purposes. This attempt, grounded on Darboux integrable functions, shall be motivated and shortly described in the sequel.

We denote $\mathcal{D}(K,\mu,\mathbb{R}^n):=\{f:K\longrightarrow\mathbb{R}^n \mid f \text{ is } \mu\text{-Darboux integrable}\}$. We denote $rc_o(K,\mu)$ the class of all non void, regular closed μ-continuity subsets of K. The *Darboux semi norm* $\|\ \|_D : \mathcal{D}(K,\mu,\mathbb{R}^n) \longrightarrow \mathbb{R}$ is defined as follows:

$$(*) \qquad \|f\|_D := \sup_{A\in rc_o(K,\mu)} \inf_{x\in A} \|f(x)\| .$$

Then $\|\ \|_D$ is a semi norm on $\mathcal{D}(K,\mu,\mathbb{R}^n)$ (i.e. the implication $\|f\|_D=0 \Rightarrow f\equiv 0$ is not necessarily fulfilled). Instead of (*) we can use the following formulas equivalently:

(**) $$\|f\|_D = \sup_{A \in rc_o(K,\mu)} \frac{1}{\mu(A)} {}^R\!\int_A \|f\| d\mu$$

(***) $$\|f\|_D = \sup\{ \|f(x)\| \mid f \text{ is continuous in } x\}.$$

Due to (***) we recognize that the μ-ess.sup norm equals $\|\ \|_D$ on $\mathcal{D}(K,\mu,\mathbb{R}^n)$. We mention that $\|\ \|_D$ can be equivalently calculated with respect to all non void, regular closed μ-continuity sets contained in an integral μ-basis field $\mathbb{B}(K,\mu)$. Moreover $\|\ \|_D$ holds analogously with respect to a dense subset S of K. If $\mathbb{R}^n$ is replaced by a compact metric space then we obtain a semi metric instead of a semi norm.

If $f \in \mathcal{D}(K,\mu,\mathbb{R}^n)$ then $\|f\|_D = 0$ iff ${}^R\!\int_K \|f\| d\mu = 0$. We define $Z\mathcal{D}(K,\mu,\mathbb{R}^n) = \{f \in \mathcal{D}(K,\mu,\mathbb{R}^n) \mid \|f\|_D = 0\}$. Let $D(K,\mu,\mathbb{R}^n) := \mathcal{D}(K,\mu,\mathbb{R}^n)/Z\mathcal{D}(K,\mu,\mathbb{R}^n)$. Then $\|\ \|_D$ induces on $D(K,\mu,\mathbb{R}^n)$ the *Darboux norm* denoted $\|\ \|_D$ too. $(D(K,\mu,\mathbb{R}^n), \|\ \|_D)$ is a *Banach space*. If $\mathbb{R}^n$ is replaced by a compact metric space - e.g. (P,δ) - then we obtain analogously a *complete metric space*. Both spaces can be considered equivalently on dense subsets of K.

Since $\|\ \|_D$ equals the μ-ess. sup norm on $D(K,\mu,\mathbb{R}^n)$ two different continuous functions are not contained in the same equivalence class of $D(K,\mu,\mathbb{R}^n)$. Moreover the continuous functions generate a closed - i.e. in particular a complete - subspace of $D(K,\mu,\mathbb{R}^n)$. Hence we denote $C(K,\mu,\mathbb{R}^n)$ *the Banach space generated by the continuous functions* in $D(K,\mu,\mathbb{R}^n)$. Obviously $C(K,\mu,\mathbb{R}^n)$ is *isometrically isomorphic* to $C(K,\mathbb{R}^n)$, the space of continuous functions. If $\mathbb{R}^n$ is replaced by a compact metric space then the above reasoning applies analogously. We obtain in this case e.g. the complete metric spaces $C(K,\mu,P)$ resp. $C(K,\mu,P \times \mathbb{R}^n_+)$. Spaces of continuous functions can be considered equivalently on dense subsets of K (via the uniformity of K). This holds too for the spaces of μ-Darboux integrable functions.

We presented the kind of equivalence classes which will serve us to transform the deterministic approach into a probabilistic one. The described finitely additive theory satisfactorily suffices many of the requirements deduced in the deterministic grounding of the model. It remains to investigate the nature and the ability of these equivalence classes with respect to our economic goals. Moreover we have to examine whether the rule A4 would be offended by a probabilistic reinterpretation of the model.

At first we want to scrutinize the nature of the above introduced equivalence classes and their aptitude for our economic goals: Let $A \subset K$ such that $A=cl(A)$ and $\mu(A)=0$. Let $f:K \to \mathbb{R}$ be continuous, and let L be a non negative, constant number. Assume that $g:K \to \mathbb{R}$ equals f on $K \setminus A$ and that $\|g(x)-f(x)\| \leq L$ for $x \in A$. Then g is a bounded modification of f on a μ^c-zero set. Due to the economic motivation from the beginning of this paragraph, g and f have to be representatives of the same equivalence class of $C(K,\mu,R)$. We show that this requirement is fulfilled:

Since the μ-Darboux norm equals the μ-ess.sup.norm on $D(K,\mu,\mathbb{R})$ we obtain $\|f-g\|_D=0$ - i.e. f and g are representatives of the same equivalence class of $C(K,\mu,\mathbb{R})$ - assumed that $g \in \mathcal{D}(K,\mu,\mathbb{R})$. The subsequent lemma together with the natural order $\geq$ on $S\mathbb{B}\mathcal{P}$ guarantee that $g \in \mathcal{D}(K,\mu,\mathbb{R})$:

<u>LEMMA II.8.</u> Let K be the compact support of the non negative Radon measure μ, and let $\mathbb{B}(K,\mu)$ be an integral μ-basis field on K. Let $\varepsilon>0$, and let $A \subset K$ be a closed μ-continuity set with $\mu(A)=0$. Then there is a $B \in \mathbb{B}(K,\mu)$ with $A \subset int(B)$ and $\mu(B)<\varepsilon$.

P r o o f: Since μ is a Radon measure, μ is regular. Hence there is an open set B_1 with $A \subset B_1$ and $\mu(B_1)<\varepsilon$. As $\mathbb{B}(K,\mu)$ contains a neighbourhood basis system of K there is for every $x \in A$ a $U(x) \in \mathbb{B}(K,\mu)$ with $x \in int(U(x))$ and $U(x) \subset B_1$. Since A is compact there are finitely many $x_1,\ldots,x_n$ in A such that $\{int(U(x_i)) \mid 1 \leq i \leq n\}$ is a finite covering of A. The union of the $U(x_i)$ yields B. Q.E.D.

The above analyzed, bounded modification of a continuous, real valued function works analogously for $\mathbb{R}_+^n$, resp. for $P\times\mathbb{R}_+^n$ valued mappings. Therefore the interpretation of continuous mappings as elements of $C(K,\mu,P\times\mathbb{R}_+)$ *suffices* the economic requirement from the beginning of this paragraph. Remember that we considered a market mapping as an equivalence class admitting modifications on μ^c-zero sets. The restriction onto bounded alterations is a direct consequence of the fact that bounded functions only are accepted as deterministic market mappings.

The above deduced, possible alteration of f on a μ^c-zero set does not completely describe the nature of an equivalence class of $C(K,\mu,\mathbb{R})$. The equivalence class of the continuous function f contains more elements of $\mathcal{D}(K,\mu,\mathbb{R})$ than one can obtain by a bounded alteration of f on one or finitely many μ^c-zero sets:

Let $\{A_i\}_{i=1}^{\infty}$ be a sequence of μ-continuity subsets of K fulfilling $\mu(\mathrm{cl}(A_i))=0$ for $i\in\mathbb{N}$ and $A_i\cap A_j=\emptyset$ for $i\neq j$. Let $\varepsilon_i>0$ and $\varepsilon_i\to 0$ for $i\to\infty$. Assume that g equals f on $K\setminus(\cup_{i=1}^{\infty}A_i)$ and that $\|g(x)-f(x)\|<\varepsilon_i$ for $x\in A_i$, $i\in\mathbb{N}$. Then $\|g-f\|_D=0$ supposing that $g\in\mathcal{D}(K,\mu,\mathbb{R})$. Due to the definition of the μ-Darboux integrability and the lemma II.8 we obtain $g\in\mathcal{D}(K,\mu,\mathbb{R})$. Hence g and f are representatives of the *same* equivalence class of $C(K,\mu,\mathbb{R})$. We mention that the same reasoning is false in general if ε_n is not tending to zero (except e.g. $\mu(\mathrm{cl}(\cup_{i=1}^{\infty}A_i))=0$).

Due to the above argument an equivalence class of $C(K,\mu,\mathbb{R})$ contains more elements of $\mathcal{D}(K,\mu,\mathbb{R})$ than it is necessary with respect to the economic motivation from the beginning of this section. This economic motive was an obvious but elementary one. If equivalence classes are introduced then we have to take care of the model consistency of the definition. An analysis of the grounding principle A4 within the probabilistic framework will demonstrate that the chosen definition of an equivalence class is consistent with our modelling:

The principle A4 reformulated with respect to the probabilistic case is as follows: (A4*) *A property of a given endowment mapping is called observable iff there exists a fixed, finite precision guaranteeing that any more precise measurement always verificates that property except on a* μ^c*-zero set* (resp. μ_S^c-zero set where S is dense in K). Remember in this context that a μ-zero set is in general not a μ^c-zero set. If we apply this principle to determine whether g is different from f or not then we obtain the following decision rule: g is different from f iff there is a $B \in \mathbb{B}(K,\mu), \mu(B) \neq 0$ - since we are considering groups only in the economic model - and an $\varepsilon > 0$ such that $\|f(x)-g(x)\| > \varepsilon$ for $x \in B$ except on a μ_c-zero set. Due to the construction of g we can never find such a B and ε. Therefore g and f have to be identified with respect to the observation principle (A4*). We conclude that the kind of equivalence classes introduced here is well motivated by (A4) resp. by (A4*). Hence the probabilistic reinterpretation used here of the model is straightforward in line with the grounding deterministic approach.

The introducing of the Jordan content-Darboux integration approach seems us to be sufficiently motivated. In particular the principles (A2)...(A5) still hold where (A4) was changed in a natural way. The principle (A1) cannot be valid here. Indeed we progressed to measure by public opinion polls which are based on methods of measurement. Therefore we can present the

<u>DEFINITION II.9.</u> Let $\overline{\mathcal{E}} := \bar{E} \times \bar{e} : \mathrm{Stone}(R_o) \longrightarrow P_{mo} \times \mathbb{R}^n_+$ be a continuous mapping. Then $\overline{\mathcal{E}}$, interpreted as an element of $C(\mathrm{Stone}(R_o), \bar{d}, P \times \mathbb{R}^n_+)$, is called an observed market mapping. The restriction $\mathcal{E}$ of $\overline{\mathcal{E}}$ onto $\mathbb{N}$ is called a market mapping. An observed market with pure competition is symbolized by $\overline{\mathcal{E}} : (\mathrm{Stone}(R_o), \overline{\mathcal{R}}_o, \bar{d}) \longrightarrow P_{mo} \times \mathbb{R}^n_+$ where $\overline{\mathcal{E}}$ is analogous as above. The restriction onto $\mathbb{N}$ yields a market with pure competition $\mathcal{E} : (\mathbb{N}, \mathcal{R}_o, d) \longrightarrow P_{mo} \times \mathbb{R}^n_+$.

Let $\mathcal{E}$ be a market mapping and let $\mathcal{F} \in \mathcal{D}(\mathbb{N}, d, P \times \mathbb{R}^n_+)$ be any representative of $\mathcal{E}$. Then $\mathcal{F}$ fulfills the stability assumption from chapter I.2.2

except on a d-zero set: If $x,y \in \mathbb{R}_+^n$, $x>y$ then there is an open set U, $(x,y) \in U \subset \mathbb{R}_+^{2n}$, with $U \subset \succ_n := F(n)$ for $n \in \mathbb{N}$ except on a d-zero set. This follows from the described nature of the equivalence classes of $C(\text{Stone}(R_o), \bar{d}, P \times \mathbb{R}_+^n)$ and the fact that the continuous representative $\bar{\mathcal{E}}$ of $\mathcal{E}$ fulfills the stability assumption.

To emphasize once more that the grounding principle A4 is not abolished but naturally modified only we want to investigate the case of a positive first endowment of commodities. In the above definition II.9 as well as in the definition II.6 a mapping $\mathcal{E} := E \times e : \mathbb{N} \longrightarrow P_{mo} \times \mathbb{R}_+^n$ with $e \equiv 0$ was not excluded. From the viewpoint of economics a market without commodities is not meaningful. The above case of the definitions can be considered as a formal notation only. Therefore we will elaborate the notion of a *positive first endowment* of commodities within our framework:

Firstly we study the case of one single commodity. Let $\bar{f}: \text{Stone}(R_o) \longrightarrow \mathbb{R}_+$ be defined by $\bar{f}(n)=1/n$ for $n \in \mathbb{N} \subset \text{Stone}(R_o)$ and $\bar{f}(x)=0$ otherwise. Then $\bar{f}$ is $\bar{d}$-Darboux integrable on $\text{Stone}(R_o)$. Let $f := \bar{f}|_{\mathbb{N}}$ the restriction of $\bar{f}$ onto $\mathbb{N}$. Then f assignes every agent $n \in \mathbb{N}$ a positive amount of the considered commodity. At a first glance one may deem f to be a suitable candidate for a market with a positive first endowment.

Observe that (i) $\bar{f}$ is not a continuous function on $\text{Stone}(R_o)$ (ii) $\bar{f}$ and the function $\bar{0}$, identical to zero on the whole space $\text{Stone}(R_o)$, are in the same equivalence class of $C(\text{Stone}(R_o), \bar{d}, \mathbb{R}_+)$ (iii) $\|\bar{f}\|_D = {}^R\!\int_{\text{Stone}(R_o)} \bar{f}\, d\bar{d} = 0$. Therefore with respect to our framework $\bar{f}$ cannot serve us as an example of a positive first endowment. This enhances the fact that there is no scale $\{[0,\varepsilon),[\varepsilon,2\varepsilon)\dots\}$ such that the strict positivity of f can be observed. Hence the simple combination of the infinite set $\mathbb{N}$ with the definition borrowed from the finite situation does *not* yield a model consistent attempt. To achieve it we apply the modified grounding principle (A4*):

Let $\bar{e}\in C(\text{Stone}(R_o),\bar{d},\mathbb{R}_+)$. Let $\bar{f}:\text{Stone}(R_o)\longrightarrow\mathbb{R}_+$ be the unique continuous representative of $\bar{e}$. Following (A4) a positive first endowment of $\bar{f}$ - resp. of the restriction $f=\bar{f}|_{\mathbb{N}}$ - is observed iff there is an $\varepsilon>0$ and a clopen set $A\subset\text{Stone}(R_o)$ with $\bar{f}(x)>\varepsilon$ for all $x\in A$ (cf. chapter I.2.2 to refer to the deterministic rule). Assume that $\bar{g}$ is *any* non continuous representative of $\bar{e}$. Then following (A4*) a positive first endowment of $\bar{g}$ is observed iff there is an $\eta>0$ and a clopen set $B\subset\text{Stone}(R_o)$ with $\bar{g}(x)>\eta$ for all $x\in B$ except on a closed $\bar{d}$-zero set (i.e. a $\bar{d}^c$-zero set). If we found A and ε for $\bar{f}$ then we can now choose B=A and $\eta=\varepsilon/2$. This is a consequence of the described nature of the representatives of $\bar{e}$.

Therefore we obtain the following definition: *A positive first endowment* of $\bar{e}$ *is observed* iff there is an $\eta>0$ and a clopen set $A\subset\text{Stone}(R_o)$ such that every representative $\bar{g}$ of $\bar{e}$ fulfills: $\bar{g}(x)>\eta$ for all $x\in A$ except on a closed $\bar{d}$-zero set (the closed $\bar{d}$-zero set depends on $\bar{g}$). This definition can be formulated equivalently for the restriction e of $\bar{e}$ onto $\mathbb{N}$.

Although the above definition is a straightforward application of (A4*) the formulation should be simplified: Remember that the $\bar{d}$-Darboux norm as well as the $\bar{d}$-Riemann integral are not dependent on the chosen representative. Therefore we obtain the following formulation, which is equivalent to the above one:

<u>DEFINITION II.10a.</u> Let $\bar{\mathcal{E}}:=\bar{E}\times\bar{e}\in C(\text{Stone}(R_o),\bar{d},P_{mo}\times\mathbb{R}^n_+)$ an observed market mapping and let $\mathcal{E}:=E\times e=\bar{\mathcal{E}}|_{\mathbb{N}}$ the related market mapping. Then $\mathcal{E}$ (resp. $\bar{\mathcal{E}}$) has a (an observed) positive first endowment with respect to the j-th commodity iff

$$\lim_{n\to\infty}\frac{1}{n}\sum_{i=1}^{n}e_j(1)>0$$

(resp. iff ${}^{R}\!\int_{\text{Stone}(R_o)}\bar{e}_j\,d\bar{d}>0$).

Verbally formulated, the above definition reads as follows: *A posi-*

tive first endowment with respect to the j-th commodity prevails in a market iff the first endowment per capita - i.e. the mean value of the first endowment - *is positive with respect to the j-th commodity*. Remember that the axiom (A4*) was our guide to reach this natural rule.

The verification of the above mentioned equivalence follows easily by inserting the unique continuous representative of $\bar{e}\in C(\mathrm{Stone}(R_o),\bar{d},\mathbb{R}^n_+)$. The equivalence described here of the introduced definition demonstrates that the Riemann Darboux integration framework is naturally adapted and consistent with the underlying deterministic base of the model.

The above definition II.10 (a) is suitable to guarantee that commodities not involved are neglected within our market model. On the other hand it is natural to erase coalitions not involved as well; i.e. coalitions whose first endowment per capita is zero with respect to each of the considered commodities. Hence we have to elaborate on the model consistent rule which guarantees that no non-void coalition is neglectable.

First the deterministic case is studied. Assume that no non-void coalition is superfluous. Then a sufficiently precise approximation of e has to affirm this statement. This implies that there must exist an $\varepsilon>0$ such that $\max\{e_j(m)\mid 1\leq j\leq n\}>\varepsilon$ for every $m\in\mathbb{N}$. This yields the following definition for the probabilistic case:

DEFINITION II.10 (b) Let $\bar{\mathcal{E}}:=\bar{E}\times\bar{e}$ be an observed market mapping and let $\mathcal{E}$ be the related market mapping. Then $\bar{\mathcal{E}}$ has a positive first endowment with respect to every non-void coalition iff there is an $\varepsilon>0$ such that

$$\max\{\bar{e}_j(x)\mid 1\leq j\leq n\}>\varepsilon$$

holds for every $x\in\mathrm{Stone}(R_o)$ except on a set $A\subset\mathrm{Stone}(R_o)$ with $\bar{d}(\mathrm{cl}(A))=0$. The restriction onto $\mathbb{N}$ yields the definition for market mappings. (c) A positive first endowment prevails in a market $\mathcal{E}$ (an observed market $\bar{\mathcal{E}}$) iff a positive first endowment prevails with respect to every commodity and with respect to every non-void coalition.

Let us shortly summarize the main results of this part: The interpretation of the deterministic, continuous model within the Riemann Darboux integration approach solved the two remaining problems: (i) the economic requirement that single agents are neglectable (ii) the requirement that only probabilistic measurements are used. Due to the intrinsic continuity character of the Riemann Darboux theory the grounding principles - except (A1) - are not abolished but reinterpreted only. Roughly speaking we can formulate the following work rule: (W1): *Solved problems within the deterministic model yield solutions for the probabilistic case by a reinterpretation according to the Darboux framework.*

We have introduced here the *static, microeconomic model of a market with pure competition* in its final form. Its motivation and economic grounding is deemed by us sufficiently detailed and clear. To reach that goal we needed the mathematical Stone space technic as well as the generalized Riemann Darboux integration approach. Both concepts are in our opinion not more complex as e.g. the abstract, idealized tool of the real numbers. But there is no doubt that both concepts are widely unfamiliar. Hence it seems to us to be necessary to present the elaborated model in a more familiar form. However we have to mention right here that this effort is possible to us at the costs of losses in the interpretation only.

4.2 THE ELEMENTARY REPRESENTATION. The closed unit interval [0,1] and the classical Riemann integration calculus are well known to every reader. Probably the compact space Stone(R_o) and the general Riemann Darboux integration theory are less familiar to non specialists. With the help of an isomorphism result we want to represent the model within the classical framework. The represented model is not always as well interpretable as the original one. At these costs non specialists are enabled in working with the model without any investment in new mathematical tools. This benefit is considered to be worthy enough to present this new, *didactically motivated* modification of the model.

The key idea for the following mathematical transformation shall be shortly illustrated: It is widely known that a homeomorphism between two compact spaces induces an isometric isomorphism between the related Banach spaces of continuous, real valued functions (or analogously: M-valued functions where M is a metric space). Similarly a *topological measure isomorphism* between two compact supports of Radon measures delivers an isometric isomorphism between the related Banach spaces of Darboux integrable functions without involving the Riemann integral.

The notion of a topological measure isomorphism was first introduced by K.KRICKEBERG [23] within the σ-additive framework. We will prefer here a finitely additive approach developed in [16] by the author. Since our interest is focused on the spaces Stone(R_o) and $[0,1]$ we are enabled to drop most of the mathematical technics. The unavoidably necessary notions and results will be shortly noted below (for the general concept and extensively elaborated proofs the reader is referred to [16]):

Let V and W be compact spaces. A continuous surjection $\Pi: V \twoheadrightarrow W$ is called *irreducible* iff for every closed set $F \subset V$ the condition $\Pi(F)=W$ implies $F=V$.

<u>PROPOSITION II.11.</u> Assume that $S:\text{Stone}(R_o) \twoheadrightarrow [0,1]$ is a continuous, irreducible surjection measure preserving on the clopen sets of Stone (R_o) - i.e. we assume that $\bar{d}(\bar{A})=\lambda(S(\bar{A}))$ holds for every $\bar{A}$, $A\in R_o$, where λ denotes the Lebesgue measure. Then S induces - via $f \mapsto f\circ S$, $f\in \mathcal{D}([0,1],\lambda,\mathbb{R}^n)$ - a Riemann integral preserving, isometric isomorphism from $D([0,1],\lambda,\mathbb{R}^n)$ onto $D(\text{Stone}(R_o),\bar{d},\mathbb{R}^n)$. Analogously S induces an isometric bijection between $D([0,1],\lambda,P)$ and $D(\text{Stone}(R_o),\bar{d},P)$.

The proof of the above proposition is a straightforward application of general results contained in [18] and in [16] chapter I § 4 and chapter II § 7. For the sake of conspicuousness we deduce the proposition explicitely. This implies necessarily that our verification adopts the terminology of [16].

P r o o f: As d is normalized and strictly positive on R_o the pair (R_o,d) is a finitely additive measure algebra. Then $(\text{Stone}(R_o),\bar{d},CO(\text{Stone}(R_o)))$ is the canonical representation space of (R_o,d). The mapping $A \mapsto \bar{A}$, $A \in R_o$ is the canonical representation where $\bar{A} \subset \text{Stone}(R_o)$ denotes the clopen set related to A. Remember that $\bar{A}=cl(A)$ if $\mathbb{N}$ is interpreted as a subset of $\text{Stone}(R_o)$.

Let $S(CO(\text{Stone}(R_o))):=\{S(\bar{A}) \mid \bar{A} \in CO(\text{Stone}(R_o))\}$. Then F denotes the field on $[0,1]$ which is generated by $S(CO(\text{Stone}(R_o)))$. We will show that $([0,1],\lambda,F)$ is a representation space of (R_o,d):

Let $\bar{A} \in CO(\text{Stone}(R_o))$. Due to [16] lemma 6 we obtain the following facts: $S(\bar{A})$ as well as $S(-\bar{A}):=S(\text{Stone}(R_o)\setminus\bar{A})$ are regular closed subsets of $[0,1]$. Moreover $\partial S(-\bar{A})=\partial S(\bar{A})=S(\bar{A})\cap S(-\bar{A})$. Since S is also measure preserving on $CO(\text{Stone}(R_o))$ we claim $\bar{d}(\bar{A})=\lambda(S(\bar{A}))=\lambda(\partial S(\bar{A}))+\lambda(S(\bar{A})\setminus\partial S(\bar{A}))$ and $\bar{d}(-\bar{A})=\lambda(\partial S(\bar{A}))+\lambda(S(-\bar{A})\setminus\partial S(\bar{A}))$. This implies $\lambda(\partial S(\bar{A}))=0$ for $1=\bar{d}(\bar{A})+\bar{d}(-\bar{A})=\lambda(S(\bar{A})\setminus \partial S(\bar{A}))+\lambda(\partial S(\bar{A}))+\lambda(S(-\bar{A})\setminus\partial S(\bar{A}))=1$. Consequently F is a field consisting of λ-continuity sets only.

In analogy to [16] lemma 7 the field F contains a neighbourhood basis system of $[0,1]$. Therefore F is a λ-basis field. This implies that $([0,1],\lambda,F)$ is a representation space of (R_o,d). A related representation φ is obtained by $A \mapsto \bar{A} \mapsto S(\bar{A})$, $A \in R_o$.

Due to the definition S is the canonical φ-projection $P(\varphi)$. Therefore $\mathfrak{A}^{CO}(\text{Stone}(R_o),\bar{d})$ and $\mathfrak{A}^{F}([0,1],\lambda)$ are measure preserving Boolean isomorphic (see [16] prop. 9 (B) and prop. 10). Due to [16] theorem 21 (remark) and [18] the proposition follows. Q.E.D.

The above proposition delivers us a theoretical instrument for a transformation of the replica model. With respect to our practical intention we need a concrete mapping S having sufficiently suitable and simple enough properties. Therefore we want to construct explicitely a special, continuous, irreducible surjection S_o which is measure preserving in the above described sense.

The construction of S_o requires some technical notations and preparations: We denoted $\{p_i\}_{i=1}^{\infty}:=\{2,3,5,7,...\}$ the indexed set of all prime numbers greater than one. Let $Q:=(q_{ij}), 1\leq i,j<\infty$ the double infinite matrix with $q_{ij}=p_i$ for $1\leq i$, $j<\infty$. We enumerate the elements of Q in analogy to G.Cantor's counting of $\mathbb{N}\times\mathbb{N}$. The resulting sequence is $(a_i)_{i=1}^{\infty}$ with $a_1=q_{11}=2$; $a_2=q_{12}=2$; $a_3=q_{21}=3$; $a_4=q_{31}=5$; $a_5=q_{22}=3$; $a_6=q_{13}=2$; $a_7=q_{14}=2$; $a_8=q_{23}=3$ etc.

In analogy to section 3.1 we denote $M[a,b](A):=\{ar-b \mid r\in A\}$ where $A\subset\mathbb{N}$, $a\in\mathbb{N}$, and $b\in\mathbb{N}\cup\{0\}$, $b<a$. This notation is generalized for the here considered case: Let $b_i\in\mathbb{N}\cup\{0\}$ $b_i<a_i$ for $i\in\mathbb{N}$. Then $M\{b_1\}:=M[a_1,b_1](\mathbb{N})$; $M\{b_1,b_2\}:=M[a_2,b_2](M[a_1,b_1](\mathbb{N}))=\{a_2(a_1r-b_1)-b_2 \mid r\in\mathbb{N}\}=\{a_2a_1r-a_2b_1-b_2 \mid r\in\mathbb{N}\}$; $M\{b_1,b_2,b_3\}:=M[a_3,b_3](M\{b_1,b_2\})$ etc. Hence $M\{b_1,...,b_n\}$ is defined for every finite sequence $\{b_i\}_{i=1}^{n}$ fulfilling the above assumption.

For the sake of clarity we explain the above notation verbally: At first we devide $\mathbb{N}$ into the odd numbers $M\{1\}$ and the even numbers $M\{0\}$. We take $M\{1\}$ and interpret it as $\mathbb{N}$. Then we divide it into the odd part $M\{1,1\}$ and the even part $M\{1,0\}$. Analogously we treat $M\{0\}$ and obtain $M\{0,1\}$ and $M\{0,0\}$. In the next step we pick out $M\{1,1\}$ and interpret it as $\mathbb{N}$. Then we divide it into three parts - since $a_3=3$ - and obtain $M\{1,1,2\}$, $M\{1,1,1\}$, $M\{1,1,0\}$ etc.

Let $z\in\mathbb{N}\cup\{\infty\}$. We denote $B(z):=\{\{b_i\}_{i=1}^{z} \mid 0\leq b_i<a_i$ and $b_i\in\mathbb{N}\cup\{0\}$ for $1\leq i<z\}$. If $n\in\mathbb{N}$ and if $\{b_i\}_{i=1}^{n}\in B(n)$ then $M\{b_1,...,b_n\}\in R_o$. We denote $\bar{M}\{b_1,...,b_n\}$ the related clopen subset of Stone(R_o), i.e. $\bar{M}\{b_1,...,b_n\}=$ cl$(M\{b_1,...,b_n\})$. Obviously $\bar{M}\{b_1,...,b_n\}\subset\bar{M}\{b_1,...,b_{n-1}\}$. If $\{b_i\}_{i=1}^{\infty}\in B(\infty)$ then we define $\bar{M}\{b_i\}_{i=1}^{\infty}:=\cap_{n=1}^{\infty}\bar{M}\{b_1,...,b_n\}$. Since Stone$(R_o)$ is compact $\bar{M}\{b_i\}_{i=1}^{\infty}$ is compact and non empty. In particular $\bar{M}\{b_i\}_{i=1}^{\infty}$ is well defined.

The class $\{\bar{M}\{b_1,\ldots,b_n\} \mid \{b_1,\ldots,b_n\} \in B(n)\}$ is a disjoint decomposition of Stone(R_o) by clopen sets. Consequently every $x \in$ Stone(R_o) is contained in exactly one compact set $\bar{M}\{b_i\}_{i=1}^{\infty}$.

Let $\{b_i\}_{i=1}^{\infty} \in B(\infty)$. We prove that $\bar{M}\{b_i\}_{i=1}^{\infty}$ contains exactly one point of Stone(R_o): Assume that $x,y \in$ Stone(R_o) with $x \neq y$ and that $x,y \in \bar{M}\{b_i\}_{i=1}^{\infty}$. Since $x \neq y$ there is an $A \in R_o$ with $x \in \bar{A}$ and $y \notin \bar{A}$. Since $A \in R_o$ there is a period s of A. Considering the decomposition of s into prime factors we recognize that there is a $m \in \mathbb{N}$ such that s divides the product $\Pi_{i=1}^{m} a_i$. Since $x \in \bar{M}\{b_i\}_{i=1}^{\infty}$ it follows that $\bar{M}\{b_i\}_{i=1}^{m} \subset \bar{A}$ due to the definition of $M\{b_i\}_{i=1}^{m}$. Hence $y \notin \bar{M}\{b_i\}_{i=1}^{\infty}$ which yields a contradiction. Therefore $\bar{M}\{b_i\}_{i=1}^{\infty}$ consists of exactly one point of Stone(R_o).

Consequently Stone(R_o) is bijective to $B(\infty)$. The set $B(\infty)$ is canonically a topological space, i.e. the product space $\Pi_{i=1}^{\infty}\{0,1,\ldots,a_i-1\}$. Due to the definition of $\bar{M}\{b_1,\ldots,b_n\}$ it follows that Stone(R_o) is *homeomorphic* to $B(\infty)$. Comparing $B(\infty)$ with the representation of Stone(R_o)= $\Pi_{n=1}^{\infty}\ \Pi_1^{\infty}\{0,1,\ldots,p_n-1\}$ from prop. II.2 we recognize that we have rearranged only the factors of the product space Stone(R_o). However this technical step was a necessary preparation for the definition of the mapping S_o.

We define S_o:Stone(R_o) $\longrightarrow [0,1]$ by $\bar{M}\{b_i\}_{i=1}^{\infty} \mapsto \sum_{i=1}^{\infty}(a_i-b_i-1)/(\Pi_{j=1}^{i} a_j)$. Observe that S_o is analogously constructed as the well known mapping from $2^{\mathbb{N}}=\{0,1\}^{\mathbb{N}}$ onto $[0,1]$ defined by $(f_i)_{i=1}^{\infty} \mapsto \sum_{i=1}^{\infty} f_i/2^i$. Obviously S_o is well defined.

<u>PROPOSITION II.12.</u> The above defined mapping S_o is a continuous, irreducible surjection from Stone(R_o) onto $[0,1]$. Moreover S_o is measure preserving on the clopen subsets of Stone(R_o) with respect to $\bar{d}$ and the Lebesgue measure λ.

P r o o f: The set $\{\bar{M}\{b_i\}_{i=1}^{\infty} \mid b_1=1\}$ is mapped by S_o onto the closed interval $[0,1/2]$. Analogously $S_o(\{\bar{M}\{b_i\}_{i=1}^{\infty} \mid b_1=0\})=[1/2,1]$, $S_o(\{\bar{M}\{b_i\}_{i=1}^{\infty} \mid b_1=1,\ b_2=1\})=[0,1/4]$ etc. Due to the product structure of $B(\infty)$ the proof is a simple exercise now. Q.E.D.

The main reason for the technical definition of S_o was not the result II.12. It was our goal to obtain a continuous irreducible surjection with additional properties suitable for our purposes. These additional properties of S_o shall be elaborated below.

We denote $\mathbb{Q}_{[0,1)}$ the *rational numbers contained in* $[0,1)$. An interval $[a,b] \subset [0,1]$ is called *rational* iff a and b are rational numbers. Then $RA([0,1],\lambda)$ denotes the *field generated by the rational subintervals* of $[0,1]$. Observe that every $A \in RA([0,1],\lambda)$ is in particular a λ-continuity set. We obtain the *Boolean algebra* $\mathfrak{A}^{RA}([0,1],\lambda) := RA([0,1],\lambda)/\{A \in RA([0,1],\lambda) \mid \lambda(A)=0\}$. The Boolean algebra $\mathfrak{A}^{RA}([0,1],\lambda)$ is a Boolean subalgebra of the Boolean algebra of all regular closed subsets of $[0,1]$. Remember that this last Boolean algebra is not a field on $[0,1]$.

<u>PROPOSITION II.13.</u> (a) Let $\mathbb{N}$ be considered as a subset of Stone (R_o). Then S_o restricted onto $\mathbb{N}$ is a bijection between $\mathbb{N}$ and the rational numbers $\mathbb{Q}_{[0,1)}$ contained in the half open unit interval $[0,1)$.
(b) S_o induces a Boolean isomorphism between the Boolean algebras CO $(\text{Stone}(R_o))$ and $\mathfrak{A}^{RA}([0,1],\lambda)$.

P r o o f: (a) If $n \in \mathbb{N} \subset \text{Stone}(R_o)$ then there is a unique sequence $\{b_i\}_{i=1}^{\infty} \in B(\infty)$ such that $n \in \bar{M}\{b_1,\dots,b_m\}$ for every $m \in \mathbb{N}$. Due to the definition of $M\{b_1,\dots,b_m\}$ there is an $mo \in \mathbb{N}$ such that n is always the first element of $M\{b_1,\dots,b_m\}$ for $m \geq mo$. This implies $a_i - b_i - 1 = 0$ for $i \geq mo$. Consequently $S_o(n)$ is a rational number and $S_o|_{\mathbb{N}}$ is injective.

Assume that $x \in [0,1)$ is a rational number, i.e. $x = p/q$ where $p,q \in \mathbb{N}$, $p<q$, and p,q have no common divisor. Since q can be decomposed into its prime factors there is a mo such that q divides $\Pi_{i=1}^{mo} a_i$. Consequently there is a sequence $\{b_i\}_{i=1}^{\infty}$, with $b_i = a_i - 1$ for $i > mo$, such that $x = \sum_{i=1}^{\infty} (a_i - b_i - 1)/\Pi_{j=1}^{i} a_j$. Then $\bar{M}\{b_i\}_{i=1}^{\infty}$ represents a natural number due to the special construction of the sequence $\{b_i\}_{i=1}^{\infty}$, i.e. $b_i = a_i - 1$ for $i > mo$. Hence S_o yields a bijection between $\mathbb{N}$ and the rational numbers contained in the half open interval [0,1).

(b) Remember that $S_o(\bar{M}\{0\})=[1/2,1]$, $S_o(\bar{M}\{1\})=[0,1/2]$ etc. It follows by induction that S_o is mapping every $\bar{A}\in CO(Stone(R_o))$, $A\neq\emptyset$ onto the union of finitely many non empty, rational intervals. Hence S_o is mapping $CO(Stone(R_o))$ into $\mathcal{A}^{RA}([0,1],\lambda)$.

On the other hand let $[a,b]\subset[0,1]$ be a non empty rational interval. Let $a=p/q$, $b=p'/q'$ such that $p<q$; $p'<q'$; $p,q,p',q'\in\mathbb{N}$. Moreover we assume that p and q as well as p' and q' possess no common divisors. Then there exists a $m_o\in\mathbb{N}$ such that q and q' divide $\Pi_{i=1}^{m_o} a_i$. This implies that [a,b] is the finite union of intervals $S_o(\bar{M}\{b_1,\ldots,b_{m_o}\})$ concerning suitable $\{b_1,\ldots,b_{m_o}\}$-combinations. Hence S_o is mapping $CO(Stone(R_o))$ onto $\mathcal{A}^{RA}([0,1],\lambda)$. Lastly it is obvious that S_o respects the Boolean operations. Q.E.D.

The deduced result II.13 establishes for the surjection S_o the first two of several special properties particularly appropriate for representing the replica model within an elementary framework. (i) *The rational numbers* $\mathbb{Q}_{[0,1)}$ contained in the half open unit interval [0,1) *are interpreted as the set of agents*. Their complement in [0,1] is seen as a counterpart of the set of non realized property combinations explained and motivated in the original model. Unfortunately we are not able to interpret all points of that complement in a similarly easy and instructive manner as we did it with respect to the Stone space tool in chapter one resp. within the replica model. This lack concerning the interpretation is attributed to the elementarily formulated definition of a large market. (ii) Concerning the probabilistic interpretation the *rational intervals are seen as the coalitions* of the elementarily represented model - resp. the intersections of rational intervals with $\mathbb{Q}_{[0,1)}$. Remember that within the probabilistic interpretation single or finitely many points can be neglected. We mention right here that in the deterministic case half open rational intervals [a,b) are chosen for representing coalitions. Then disjoint coalitions are related to disjoint represented coalitions. Unfortunately the elementary representation of the deterministic model requires still some more mathematical tricks. They shall be elaborated below as well as the meaning of represented market mappings.

In the Replica model market mappings are characterized by special continuous functions defined on Stone(R_o), resp. by the corresponding equivalence classes in $D(\text{Stone}(R_o),\bar{d},P\times\mathbb{R}^n_+)$. Hence we are faced with the problem of describing those equivalence classes in $D([0,1],\lambda,P\times\mathbb{R}^n_+)$ which are related to market mappings defined on Stone(R_o). Moreover, if possible, we should try to detect suitable, deterministic functions defined on $[0,1]$ which generate those represented market mappings. This problem under consideration is solved in an indirect way elaborated below.

We consider the set $w([0,1]):=\{0,1\}\cup\{a\in[0,1]\mid a \text{ is irrational}\}\cup\{q^+:=(q,0)\mid q\in(0,1)\cap\mathbb{Q}\}\cup\{q^-:=(0,q)\mid q\in(0,1)\cap\mathbb{Q}\}$; i.e. we consider the unit interval where every rational number, except the boundary points, is duplicated. On $w([0,1])$ we use the familiar order $\leq$ completed by $q^-\leq q^+$ for $q\in(0,1)\cap\mathbb{Q}$. Let us call $w([0,1])$ *the* w-*unit interval.*

Let $\{q_i\}_{i=1}^{\infty}$ be any enumeration of the rational numbers $(0,1)\cap\mathbb{Q}$. A metric d_w is obtained for the w-unit interval as follows: Let $\{c_i\}_{i\in\mathbb{N}}\subset\mathbb{R}_+\setminus\{0\}$ be a sequence with $\sum_{i=1}^{\infty}c_i=1$. If $a,b\in w([0,1])$ then $d_w(a,b):=|a-b|+\sum_{i\in I(a,b)}c_i$ where $I(a,b):=\{i\in\mathbb{N}\mid a\leq q_i^-, q_i^+\leq b \text{ or } b\leq q_i^-, q_i^+\leq a\}$. One easily verifies that $(w([0,1]), d_w)$ is a metric space fulfilling the second countability axiom. Therefore the w-unit interval is compact for one easily verifies its sequential compactness. Consequently, *the* w-*unit interval* $(w([0,1]),d_w)$ *is a compact metric space.*

<u>LEMMA II.14.</u> Stone(R_o) is homeomorphic to the w-unit interval.

P r o o f: Remember that Stone(R_o) can be identified with $B(\infty)$. Then we define: (i) $S_o(w)(\{b_i\}_{i=1}^{\infty})=q_+$ iff $b_i=a_i-1$ for almost all $i\in\mathbb{N}$ and $S_o(\{b_i\}_{i=1}^{\infty})=q\in(0,1)\cap\mathbb{Q}$. (ii) $S_o(w)(\{b_i\}_{i=1}^{\infty})=q_-$ iff $a_i<a_i-1$ for infinitely many $i\in\mathbb{N}$ and $S_o(\{b_i\}_{i=1}^{\infty})=q\in(0,1)\cap\mathbb{Q}$. (iii) $S_o(w)\equiv S_o$ in all

other cases. Due to the canonical definition of S_o it turns out that $S_o(w)$ is the proposed homeomorphism. Q.E.D.

By the previous lemma II.14 it follows that $C(\mathrm{Stone}(R_o),M)$ *is isometrically bijective to* $C(w([0,1]),M)$ provided $M=P,\mathbb{R}_+^n$, or $M=P\times\mathbb{R}_+^n$. If $M=\mathbb{R}$ or $M=\mathbb{R}^n$ then we obtain analogously an *isometric isomorphism*. In this connexion remember that $n\in\mathbb{N}\subset\mathrm{Stone}(R_o)$ is bijectively mapped by $S_o(w)$ onto $\{0\}\cup\{q^+ \mid q\in(0,1)\}$. Hence not only deterministic observed markets but even deterministic markets can be well identified within this newly introduced framework.

Declaring $\lambda(q_i^+)=\lambda(q_i^-):=0$ for $i\in\mathbb{N}$ we can *extend* the Lebesgue measure λ onto the w-unit interval $w([0,1])$. Obviously the homeomorphism $S_o(w)$ yields automatically a (σ-additive) measure isomorphism. In particular $S_o(w)$ *preserves the Riemann integral*. Moreover it follows that $C(\mathrm{Stone}(R_o),\bar{d},M)$ *is isometrically bijective to* $C(w([0,1]),\lambda,M)$ if $M=\mathbb{R}_+^n,P$ or $=P\times\mathbb{R}_+^n$. In the cases $M=\mathbb{R}$ and $M=\mathbb{R}^n$ we obtain analogously an *isometric isomorphism*.

The reader waiting for the elementary representation of the Replica model may get impatient at this again new compact space $w([0,1])$. The advantage of the w-unit interval is caused by its adjacency to the unit interval as well as by the fact that we obtained not only a true picture of the Darboux integrable functions but also of the continuous mappings. This benefit will enable us to detect easily a suitable class of deterministic functions generating that subspace of $D([0,1],\lambda,P\times\mathbb{R}_+^n)$ which corresponds to $C(\mathrm{Stone}(R_o),\bar{d},P\times\mathbb{R}_+^n)$ with respect to S_o. Indeed we are prepared now to undertake the final step on our way to an elementary representation of the replica model.

Once more let M be one of the following metric resp. Banach spaces: $\mathbb{R},\mathbb{R}^n,P,P\times\mathbb{R}_+^n$. Let f be a bounded, M-valued function defined on the unit interval [0,1]. Then f is called *right continuous* iff $\lim_{x\downarrow a}f(x)=f(a)$ holds for every $a\in[0,1)$. The function f is said to be *left continu-*

ously extensible iff $\lim_{x\uparrow a} f(x)$ exists for every $a\in(0,1]$. If $0\leq a<b\leq 1$ then the characteristic function $\chi_{[a,b)}$ of the half open interval $[a,b)$ is a right continuous, left continuously extensible function.

We denote $REP([0,1],M):=\{f:[0,1]\to M \mid f$ is bounded, right continuous, left continuously extensible, and continuous in every $x\in[0,1]\setminus((0,1)\cap\mathbb{Q})\}$. On $REP([0,1],M)$ we consider the sup metric resp. the sup norm. Observe that a convergent sequence concerning the sup metric preserves the continuity character with respect to the limit function. Therefore REP ([0,1],M) *is a complete metric space*. If $M=\mathbb{R}$ or $M=\mathbb{R}^n$ then $REP([0,1],M)$ is a *Banach space*. Due to its special definition one verifies immediately that $REP([0,1],M)$ *is isometrically bijective* - resp. isometrically isomorphic - *to* $C(w([0,1]),M)$. Consequently $REP([0,1],M)$ *is isometrically bijective* - resp. isometrically isomorphic - *to* $C(\mathrm{Stone}(R_o),M)$.

We define the canonical surjection $S_w:w([0,1])\longrightarrow[0,1]$ by $q^+,q^-\longmapsto q$ for $q\in(0,1)\cap\mathbb{Q}$ and by $S_w(x)=x$ for all other points in $[0,1]$. Then S_w *is a continuous, irreducible surjection.* Moreover $S_o=S_w\circ S_o(w)$.

By prop. II.11 it follows that $D(w([0,1]),\lambda,M)$ is isometrically bijective - resp. isometrically isomorphic if $M=\mathbb{R}$ or $M=\mathbb{R}^n$ - to $D([0,1],\lambda,M)$. We denote $REP([0,1],\lambda,M)$ the space generated by $REP([0,1],M)$ in $D([0,1],\lambda,M)$. Then $REP([0,1],\lambda,M)$ *is isometrically bijective* (isomorphic) *to* $C(w([0,1]),\lambda,M)$. Consequently $REP([0,1],\lambda,M)$ *is isometrically bijective* (isomorphic) *to* $C(\mathrm{Stone}(R_o),\bar{d},M)$.

Observe that the above introduced definitions of $REP([0,1],M)$ as well as of $REP([0,1],\lambda,M)$ require elementary notions only, i.e. continuity, completeness and the classical Riemann integration. Indeed, after a series of mathematical tricks we are prepared now to formulate the elementary representation of the Replica model. (We have to mention that the last trick is studied in a more general context in T.E.ARMSTRONG [4], H.D.BRUNK [7], and Z.SEMADENI [34]).

<u>DEFINITION II.15.</u> We consider the unit interval [0,1] endowed with

the Lebesgue measure λ. (i) The rational numbers contained in $[0,1)$ are called represented agents. All other points of the unit interval are corresponding to non realized property combinations. The half open intervals $[a,b)$ having rational endpoints are called represented observed coalitions. Their intersections with the represented agents are called represented coalitions. (ii) A mapping $\bar{\mathcal{E}}:=\bar{E}\times\bar{e}\in REP([0,1],P\times\mathbb{R}^n_+):=\{\bar{\mathcal{E}}:[0,1]\longrightarrow P\times\mathbb{R}^n_+ \mid \bar{\mathcal{E}}$ is bounded, right continuous, left continuously extensible, and can possess discontinuities in rational points of $[0,1)$ only$\}$ is called a represented, deterministic, observed endowment mapping. If $cl(\bar{E}([0,1]))\subset P_{mo}$ then $\bar{\mathcal{E}}$ is called a represented, deterministic, observed market mapping. The restriction $\mathcal{E}$ of $\bar{\mathcal{E}}$ onto the rational numbers in $[0,1)$ is a represented, deterministic endowment - resp. market - mapping. (iii) $\bar{\mathcal{E}}$, interpreted as an element of the space of Darboux equivalence classes, i.e. equivalence classes with respect to the essential sup norm, is called a represented, observed endowment - resp. market - mapping. The restriction onto the rational numbers in $[0,1)$ is the represented endowment - resp. market - mapping. (iv) A positive first endowment prevails in $\bar{\mathcal{E}}$ iff the endowment per capita is positive with respect to every commodity, i.e.

$$\int_{[0,1]}\bar{e}d\lambda \gg 0,$$

and with respect to every non void coalition, i.e. there is an $\varepsilon>0$ with

$$\max\{\bar{e}_j(x) \mid 1\leqq j\leqq n\}>\varepsilon$$

for every $x\in[0,1]$ except on a Jordan zero set (i.e. a subset $A\subset[0,1]$ with $\lambda(cl(A))=0$). By J.v.Neumann's rearrangement theorem the integral can be calculated by a limit of average means over the represented agents. (v) Let $\bar{\mathcal{E}}$ be a represented, observed market mapping. Then a represented, observed market with pure competition is symbolized by $\bar{\mathcal{E}}$:

$([0,1],\mathcal{R}A([0,1]),\lambda^c) \longrightarrow P_{mo}\times\mathbb{R}^n_+$. Here $\mathcal{R}A([0,1])$ denotes the field generated by the rational intervals and the Jordan-zero-sets. λ^c denotes the Jordan content. A represented market with pure competition is obtained by the canonical restriction onto the rational numbers.

In the above definition II.15 the realization of the monotonicity condition for market mappings, derived in chapter I.2.2, cannot be guaranteed in such an elegant manner as this was possible within the deterministic replica model. Indeed, although $C(\text{Stone}(R_o),P\times\mathbb{R}^n_+)$ is isometric to $REP([0,1],P\times\mathbb{R}^n_+)$, the range of a mapping in $C(\text{Stone}(R_o),P\times\mathbb{R}^n_+)$ is in general only the closure of the related mapping defined on the unit interval. Therefore the explicit monotonicity condition for market mappings cannot be ignored here. Its justification and motivation naturally remain the same as before. To emphasize nevertheless the adjacency between $REP([0,1],P\times\mathbb{R}^n_+)$ and spaces of continuous functions we remind the reader of the subsequent, reformulated approximation property:

<u>REMARK II.16.</u> Let $\overline{\mathcal{E}}\in REP([0,1],P\times\mathbb{R}^n_+)$ and let $\varepsilon>0$. Then there are finitely many rational numbers $r_1,\dots,r_{n(\varepsilon)}$ contained in $(0,1)$ such that the diameter dia $(\overline{\mathcal{E}}([r_i,r_{i+1})))<\varepsilon$ for $i=0,\dots,n(\varepsilon)+1$ where $r_o=0$ and $r_{n(\varepsilon)+1}=1$. Hence $\overline{\mathcal{E}}$ can be approximated by simple functions contained in $REP([0,1],P\times\mathbb{R}^n_+)$.

P r o o f: Remember that $REP([0,1],P\times\mathbb{R}^n_+)$ is isometric to $C(w([0,1]),P\times\mathbb{R}^n_+)$. As $w([0,1])$ is homeomorphic to $\text{Stone}(R_o)$ the space $w([0,1])$ is zero dimensional. Hence proposition II.5 yields the proof. Q.E.D.

By a series of mathematical tricks we reached our goal and could present the elementarily formulated definition of a large market with pure competition. The resulting definition is not as perfectly interpretable and elegant as the original definition of the replica model. At these costs we hope that a reader, familiar with the principal grounding of the model, is enabled to use it without essential investment in unfamiliar mathematical framework. However we cannot conceal the fact that the picture of a set of agents, threaded on the real line, seems to us to be not directly natural and less stimulating to a microeconomic analysis. Indeed the above definition is actually a concession to the prevailing mathematical education grounded quite often in historical requirements of classical physics.

4.3 CONCLUSION. In the third part of this article we installed the deterministic, concrete model of a large market with pure competition. Since the measurement by public opinion polls as well as the problem of pure competition intrinsicly required a probabilistic consideration a reinterpretation of the deterministic replica model was necessary and unavoidable.

The measurement by probabilistic methods and the neglecting of single agents within the status of pure competition caused a reinterpretation of the meaning of coalitions as well as of market mappings. Both had to be seen as equivalence classes determined by the system of neglectible sets, i.e. by the system of zero influence sets. Scrutinizing the requirements of our modelling at the system of zero influence sets we were forced to abandon the Lebesgue measure theory as a mathematical tool for introducing the probabilistic character into the replica model.

Instead of the Lebesgue measure theory the generalized Jordan content-Riemann Darboux integration approach was chosen to obtain the probabilistic reinterpretation of the deterministic replica model. Market mappings were understood as elements of the complete space of Darboux equivalence classes and coalitions were seen as deterministic coalitions variated by Jordan zero sets. Although we motivated every detail of the new probabilistic replica model the result turned out to be a simple *reinterpretation* of the hard elaborated, deterministic model. In particular the model-axioms (A2), (A3), and (A5) remained unaltered whereas

(A4) was only canonically changed to (A4*). The linkage between the deterministic and the probabilistic model was even so firm that the work-rule (W1) could be stated: Solved problems within the deterministic model yield solutions for the probabilistic case by a reinterpretation according to the Darboux framework. The result of this probabilistic reinterpretation was the carefully explained and motivated definition of a *static, microeconomic market with pure competition.*

Although the developed model of a market with pure competition could be justified satisfactorily there remained no doubt that its appreciation provided the familiarity with the abstract Stone space tool and the generalized Riemann Darboux approach. Therefore the didactical reason to present the model within an elementary framework could not be neglected. By a series of mathematical tricks the *didactically motivated, elementary formulation* of a market with pure competition could be obtained at the costs of slight losses within the economic interpretability.

CHAPTER III. CORE AND WALRAS ALLOCATIONS

The major motivation of earlier studies on large markets with pure competition was always the aspiration for a well interpretable verification of F.Y.Edgeworth's conjecture concerning the Core-Walras equivalence. Consequently it is our natural desire to investigate this classical conjecture within the model installed here and its intrinsic logic. Since our approach differs from other contributions to this theme we have to deduce the necessary notions before a verification of the equivalence result can be probed.

5. THE DEFINITION OF THE CORE

In the case that the reader is informed about earlier approaches concerning F.Y.Edgeworth's conjecture he may await that a direct reinterpretation of former definitions within the Riemann Darboux framework delivers the notion announced in the title. Instead of such an adoption we will develop, step by step, the necessarily idealized and abstract definitions. The natural reason is our duty to take care of the compatibility of the new definitions with the grounding observability approach as well as to explain the connection with the real situation given by the finite, discrete market. Indeed our analysis can be based only on the scanning-completion picture, developed in the first two chapters, and the finite case provided that a model-consistent definition should be obtained.

5.1 ATTAINABLE ALLOCATIONS. The starting point of our considerations is once again the finite case. Let $N=\{1,...,n\}$ and let $\mathfrak{F}:=F\times f:(N,\mathbb{P}(N),1_n)\to P_{mo}\times\mathbb{R}_+^n$ a finite market. Then an allocation $g:N\to\mathbb{R}_+^n$

is called *attainable* with respect to $\mathfrak{E}$ iff $\sum_{i=1}^{n} g(i) \leq \sum_{i=1}^{n} f(i)$. An attainable allocation with respect to $\mathfrak{E}$ is seen as a possible redistribution of the commodities without increasing the whole amount of every one of the considered commodities. Hence the set of all attainable allocations with respect to $\mathfrak{E}$ is the set of all alternative redistributions of commodities which can be realized so far as preferences and tastes of the agents are neglected.

Although the introduced notion is an elementary one we should keep in mind that an attainable allocation as well as the set of all attainable allocations is an economic abstraction. Indeed with respect to markets in reality - i.e. finite, discrete markets - the set of possible alternatives is much smaller. Since the real numbers are well interpretable as an idealization derived from the rational numbers with the help of the natural Euclidean metric, no serious difficulties arise for a clear understanding and interpretation.

Next we study the case of pure competition. Assume that $\bar{\mathfrak{E}} := \bar{E} \times \bar{e}: \text{Stone}(R_o) \rightarrow P_{mo} \times \mathbb{R}_+^n$ is an observed market mapping. An *observed allocation* is a continuous mapping $\bar{f}: \text{Stone}(R_o) \rightarrow \mathbb{R}_+^n$ interpreted within the Darboux framework. If $\bar{f}$ additionally fulfills

$$\int_{\text{Stone}(R_o)} \bar{f} d\bar{d} \leq \int_{\text{Stone}(R_o)} \bar{e} d\bar{d}$$

then $\bar{f}$ is called an *attainable observed allocation* with respect to the observed market $\bar{\mathfrak{E}}$ with pure competition. The restriction f of $\bar{f}$ onto $\mathbb{N} \subset \text{Stone}(R_o)$ yields an *allocation* - resp. an *attainable allocation* - with respect to the market $\mathfrak{E}$ with pure competition. In this connection remember that the above integral-condition can be formulated equivalently as follows:

$$\lim_{n\to\infty} (1/n) \sum_{i=1}^{n} f(i) \leq \lim_{n\to\infty} (1/n) \sum_{i=1}^{n} e(i).$$

In the finite situation attainable allocations are characterized by a non increasing whole amount of every considered commodity, i.e. by an absolute kind of measurement. As the status of pure competition is an extreme microeconomic idealization it was unavoidable to use in that case a relative variable, i.e. *an allocation is attainable iff the amount per capita* of every considered commodity *fails to grow*. Observe that this microeconomic condition automatically erases the problem that not single mappings but equivalence classes of mappings are considered in the case of pure competition.

In analogy to the finite situation an attainable allocation is considered as a possible redistribution of commodities without adding any measurable amount of goods. Then the set of all attainable allocations is the set of all alternative redistributions of commodities where prodigality is allowed. Since the notions considered here are based on the replica model a more explicit analysis of the above analogy seems to us to be useful:

By section II.3.1 a finite market $\mathfrak{F}$ defined on $N=\{1,...,n\}$ yields by an infinite replication an idealized market $\mathfrak{E}$ on $\mathbb{N}$ with the period n. So long as the set of coalitions is restricted to BG(n) both markets $\mathfrak{E}$ and $\mathfrak{F}$ are equivalent. This equivalence implies that $\mathfrak{E}$ is not a market with pure competition. Assume that f' is an attainable allocation with respect to $\mathfrak{F}$. Then f' delivers by an infinite replication an attainable allocation e' with respect to $\mathfrak{E}$ where e' possesses the period n too. All attainable allocations with respect to $\mathfrak{E}$ and BG(n) are obtained in this manner.

If the status of pure competition is more precisely approximated - e.g. by BG(mn), $m\in\mathbb{N}$, $m>1$, instead of BG(n) - then the set of attainable allocations with respect to $\mathfrak{E}$ is enlarged. It includes all allocations with the period mn fulfilling the summability condition of an unchanged per-capita amount of goods.

The status of pure competition is reached within the deterministic framework if $R_o = \bigcup_{n\in\mathbb{N}} BG(n)$ symbolizes the set of all coalitions. Then every continuous, $\mathbb{R}^n_+$-valued function on Stone(R_o) - resp. its restriction onto $\mathbb{N}$ - respecting the per-capita restriction is an attainable allocation. Therefore the set of all attainable allocations within the case of pure competition includes all attainable allocations arising in any approximating stage of pure competition. Moreover the set of all attainable allocations within the pure case turns out to be the closure - and therefore the completion - of the collection of all attainable allocations arising in any stage of an approximated pure competition.

In the finite case the abstract, $\mathbb{R}^n_+$-valued, attainable allocations include all attainable allocations arising in related real markets, i.e. discrete markets having the same first endowment of commodities. Moreover the set of all $\mathbb{R}^n_+$-valued, attainable allocations is the completion of the collection of all attainable allocations arising in related realizations. Consequently the step from the real to the $\mathbb{R}^n_+$-valued, attainable allocations concerning the finite case is an exact analogy to its counterpart within the deterministic replica model. Due to the firm connection between the deterministic and the probabilistic replica models the application of the work rule (W1) changes neither the interpretation elaborated here nor the analogy between the sets of alternative allocations.

By the above explanation one recognizes the set introduced here of attainable allocations as a direct combination of the definition for the finite case and the inherent principles of the replica model.

We finish this section by the elementarily formulated definition of an attainable allocation: Let $\overline{\mathcal{E}} := \bar{E}\times\bar{e}: [0,1] \to P_{mo}\times\mathbb{R}^n_+$ be a represented observed market mapping. A bounded, right continuous, left continuously extensible mapping $\bar{e}': [0,1] \to \mathbb{R}^n_+$ having rational discontinuities only is called a deterministic, *represented, observed allocation*. A deterministic, represented, observed allocation e' is called *attainable* with respect to $\overline{\mathcal{E}}$ iff

$$\int_{[0,1]}\bar{e}'d\lambda \leq \int_{[0,1]}\bar{e}d\lambda .$$

The restriction of $\bar{e}'$ onto $\mathbb{Q}\cap[0,1)$ is called a deterministic, *represented allocation*, resp. a deterministic, *represented, attainable allocation*, with respect to the represented market $\mathcal{E} = \overline{\mathcal{E}}|_{\mathbb{Q}\cap[0,1)}$. Interpreting the allocations as Darboux equivalence classes yields the related notions within the probabilistic model.

5.2 PREFERRED ALLOCATIONS. In the previous part we introduced the set of all possible redistributions of commodities of a market without paying attention to tastes or preferences of the members of that market. Here a model consistent concept is developed describing a way of comparing two different allocations in conformity with the preferences of a group of agents. The economic notion of the Core will be based on this rule for a comparison similarly as the meaning of a monotonic endowment mapping is based on the model consistent notion of a property of a market.

Keeping in mind the meaning of our work-rule (W1) we elaborate at first a model consistent comparison between two allocations within the deterministic case of the model. The thread of the development of our definition will be the grounding axiom (A4) resp. later on its canonical translation (A4*) for the probabilistic situation.

Assume that $\succ \in P_{mo}$ is a monotone preference, that $x,y\in\mathbb{R}^n_+$, and that $\varepsilon\in\mathbb{R}, \varepsilon>0$. Then x is called *ε-preferred to* y *by* $\succ$ iff $U_\varepsilon((x,y)):=\{z\in\mathbb{R}^{2n}_+ \mid \|z-(x,y)\|<\varepsilon\} \subset \succ$. If $x \succ y$ then there exists an $\varepsilon_o>0$ such that x is ε-preferred to y by $\succ$ for every ε with $0<\varepsilon\leq\varepsilon_o$. This follows because $\succ$ is an open subset of $\mathbb{R}^{2n}_+$.

Let $\mathcal{E} := E\times e:(\mathbb{N}, R_o, d) \to P_{mo}\times\mathbb{R}^n_+$ a deterministic market with pure competition. Let f,g be *deterministic allocations*, i.e. we suppose that

f,g possess continuous extensions $\bar{f},\bar{g}$ onto Stone(R_o) and that f,g are not interpreted within the Darboux framework. Let $A\in R_o, A\neq\emptyset$ and let $\varepsilon\in\mathbb{R}$, $\varepsilon>0$. Then f is called *ε-preferred to* g *by* A - denoted: $f\varepsilon\text{-}\succ_A g$ - iff $f(n)$ is ε-preferred to $g(n)$ by $E(n)$ for every $n\in A$. Then f is *preferred to* g *by* A - denoted: $f\succ_A g$ - iff there exists an $\varepsilon>0$ such that f is ε-preferred to g by A.

In the case of an observed, deterministic market we apply the above notions analogously: $\bar{f}$ is ε-preferred to $\bar{g}$ by $\bar{A}$ iff $\bar{f}(x)$ is ε-preferred to $\bar{g}(x)$ by $\bar{E}(x)$ for every $x\in\bar{A}=cl(A)$. Then $\bar{f}$ is preferred to $\bar{g}$ by $\bar{A}$ iff there is an $\varepsilon>0$ such that $\bar{f}$ is ε-preferred to $\bar{g}$ by $\bar{A}$.The notations are applied analogously too.

PROPOSITION III.1. Let $\mathcal{E}:=E\times e:(\mathbb{N},R_o,d)\rightarrow P_{mo}\times\mathbb{R}^n_+$ a deterministic market with pure competition. Let f,g be deterministic allocations and let $A\in R_o, A\neq\emptyset$. Then $\bar{f}$ is preferred to $\bar{g}$ by $\bar{A}$ iff f is preferred to g by A.

P r o o f: It suffices to prove the if direction. Assume that f is preferred to g by A. Then there exists an $\eta>0$ such that f is η-preferred to g by A.

Let f',g' be R_o-simple functions with $\|f-f'\|:=\sup\{\|f(n)-f'(n)\| \mid n\in\mathbb{N}\}<\eta/9$ and $\|g-g'\|<\eta/9$. Let E' be a R_o-simple, P-valued mapping with $\sup\{\delta(E(n),E'(n))\mid n\in\mathbb{N}\}<\eta/9$. As f',g', and E' are R_o-simple they extend continuously onto Stone(R_o). Moreover $\|\bar{f}-\bar{f}'\|\leq\eta/9$, $\|\bar{g}-\bar{g}'\|\leq\eta/9$, and $\sup\{\delta(\bar{E}(x),\bar{E}'(x))\mid x\in\text{Stone}(R_o)\}\leq\eta/9$.

Since the distance between f,f', resp. g,g', and E,E', is less than $\eta/9$ it follows that f' is η_1-preferred to g' by A with respect to E' where $\eta_1=\eta-3\eta/9=2\eta/3$. Then $\bar{f}'$ is η_1-preferred to $\bar{g}'$ by $\bar{A}=cl(A)$ with respect to $\bar{E}'$ because f',g', and E' are R_o-simple. As the distance between $\bar{f},\bar{f}'$, resp. $\bar{g},\bar{g}'$, and $\bar{E},\bar{E}'$, is less or equal $\eta/9$ it follows that $\bar{f}$ is ε-preferred to $\bar{g}$ by $\bar{A}$ with respect to $\bar{E}$ where $0<\varepsilon<2\eta/3-3\eta/9=\eta/3$. Q.E.D.

Although the above proposition can be seen as a sufficient hint at the model-consistency of the new notions we want to justify them explicitely:

Let $N=\{1,\dots,n\}$ a finite set of agents and let $\mathcal{E}:=F\times f:(N,\mathbb{P}(N),l_n) \rightarrow P_{mo}\times\mathbb{R}^n_+$ a finite market. Let $g,h:N\rightarrow\mathbb{R}^n_+$ be two allocations and let $A\subset N$, $A\neq\emptyset$. Then g is preferred to h by A concerning $\mathcal{E}$ iff g(n) is preferred to h(n) with respect to F(n) for every $n\in A$. Indeed, we deduced that g(n) is preferred to h(n) by F(n) iff there is an $\varepsilon_n\in\mathbb{R}$, $\varepsilon_n>0$ with $g(n)\varepsilon_n\text{-}\succ_n h(n)$. Hence $g\succ_A h$ iff $g\varepsilon\text{-}\succ_A h$ where $\varepsilon=\min\{\varepsilon_n|n\in A\}$. Therefore our concept of coalition-preference coincides with the classical coalitional-preference in the case of finitely many agents.

Within the situation of a countably infinite set $\mathbb{N}$ of agents we recognize the definition of the coalitional-ε-preference as a direct application of the grounding principle (A4): Due to the intrinsic logic of the proof of prop. III.1 the coalitional-ε-preference can always be verified by every approximation finer than a prescribed, finite precision. Hence the notion of the coalitional-ε-preference is straightforward in line with our grounding, deterministic modelling rules.

The notion of coalitional-preference is deduced by collecting and including all perception of comparison which we are able to realize in any stage of coalitional-ε-preference. As our model is interpreted as the idealized superstructure of all approximative levels of pure competition the final resulting definition necessarily has to regard and to contain every comparison-rule applicable in any stage of precision. The obtained definition is consistent with the axiom (A4): If f is preferred to g by A then there is a fixed, finite precision guaranteeing that $f\succ_A g$ is recognized by every finer approximation. We conclude that the elaborated definitions are well justified concerning the consistency of the deterministic model. Consequently the canonical use of the work rule (W1) yields the definitions within the probabilistic model:

<u>DEFINITION III.2.</u> Let $\bar{\mathcal{E}}:=\bar{E}\times\bar{e}:(\mathrm{Stone}(R_o),\bar{\mathcal{R}}_o,\bar{d})\rightarrow P_{mo}\times\mathbb{R}^n_+$ an ob-

served market with pure competition. Let $\bar{f},\bar{g}\in C(\text{Stone}(R_o),\bar{d},\mathbb{R}^n_+)$ be observed allocations and let $\bar{A}\in CO(\text{Stone}(R_o))$ with $\bar{A}\neq\emptyset$. (i) Let $\varepsilon\in\mathbb{R},\varepsilon>0$. Then $\bar{f}$ is ε-preferred to $\bar{g}$ by $\bar{A}$ with respect to $\bar{\mathcal{E}}$ - denoted $\bar{f}\varepsilon\text{-}\succ_{\bar{A}}\bar{g}$ - iff the following holds: Let $\bar{f}',\bar{g}'\in\mathcal{D}(\text{Stone}(R_o),\bar{d},\mathbb{R}^n_+),\bar{E}'\in\mathcal{D}(\text{Stone}(R_o),\bar{d},P_{mo})$ be any representatives of $\bar{f},\bar{g}$ resp. $\bar{E}$. Let $\bar{A}'\in\bar{\mathcal{R}}_o$ be any representative of $\bar{A}$, i.e. int cl$(\bar{A}')=\bar{A}$. Then $\bar{f}'(x)$ is ε-preferred to $\bar{g}'(x)$ by $\bar{E}'(x)$ for every $x\in\bar{A}'$ except on a subset $N\subset\bar{A}'$ with $\bar{d}(\text{cl}(N))=0$. Here N may depend on $\bar{f}',\bar{g}',\bar{E}'$, and $\bar{A}'$. (ii) $\bar{f}$ is preferred to $\bar{g}$ by $\bar{A}$ with respect to $\bar{E}$ - denoted $\bar{f}\succ_{\bar{A}}\bar{g}$ - iff there is an $\varepsilon\in\mathbb{R},\varepsilon>0$ such that $\bar{f}\varepsilon\text{-}\succ_{\bar{A}}\bar{g}$. (iii) By restricting $\bar{f},\bar{g},\bar{E}$ onto $\mathbb{N}$ and by considering $A=\bar{A}\cap\mathbb{N}$ we obtain the related definitions for markets with pure competition.

<u>COROLLARY III.3.</u> Let $\bar{\mathcal{E}}:=\bar{E}\times\bar{e}:(\text{Stone}(R_o),\bar{\mathcal{R}}_o,\bar{d})\to P_{mo}\times\mathbb{R}^n_+$ an observed market with pure competition and let $\mathcal{E}=\bar{\mathcal{E}}|_{\mathbb{N}}$ the related market with pure competition. Let $\bar{f},\bar{g}\in C(\text{Stone}(R_o),\bar{d},\mathbb{R}^n_+)$, and let $\bar{A}\in CO(\text{Stone}(R_o))$, $\bar{A}\neq\emptyset$. Let f,g be the related restrictions onto $\mathbb{N}$ and $A=\bar{A}\cap\mathbb{N}$. Then $\bar{f}\succ_{\bar{A}}\bar{g}$ with respect to $\bar{\mathcal{E}}$ iff $f\succ_A g$ with respect to $\mathcal{E}$.

We complete our notions by the formulation of the rules of comparison within the elementarily represented model. Let $\bar{\mathcal{E}}:=\bar{E}\times\bar{e}:([0,1],\mathcal{R}A([0,1]),\lambda^c)\to P_{mo}\times\mathbb{R}^n_+$ be a represented, observed market with pure competition. Let $\bar{f},\bar{g}$ be represented, observed allocations, and let $\bar{A}\in\mathcal{R}A([0,1]),\bar{A}\neq\emptyset$. (i) Let $\varepsilon\in\mathbb{R},\varepsilon>0$. Then $\bar{f}$ *is* ε*-preferred to* $\bar{g}$ *by* $\bar{A}$ with respect to $\bar{\mathcal{E}}$ - denoted $\bar{f}\varepsilon\text{-}\succ_{\bar{A}}\bar{g}$ - iff the following holds: Let $\bar{f}',\bar{g}'$, and $\bar{E}'$ be any representatives of $\bar{f},\bar{g}$, and $\bar{E}$. Then $\bar{f}'(x)$ is ε-preferred to $\bar{g}'(x)$ by $\bar{E}'(x)$ for every $x\in\bar{A}$ except on a set N having a zero Jordan content. Here N may depend on $\bar{f}',\bar{g}',\bar{E}'$, and $\bar{A}$. (ii) $\bar{f}$ *is preferred to* $\bar{g}$ *by* $\bar{A}$ with respect to $\bar{E}$ - denoted $\bar{f}\succ_{\bar{A}}\bar{g}$ - iff there is an $\varepsilon\in\mathbb{R},\varepsilon>0$ such that $\bar{f}\varepsilon\text{-}\succ_{\bar{A}}\bar{g}$. (iii) By restricting $\bar{f},\bar{g},\bar{E}$ onto $\mathbb{Q}\cap[0,1)$ and by considering $A=\bar{A}\cap\mathbb{Q}\cap[0,1)$ we obtain the related definitions for represented markets with pure competition.

5.3 UNACCEPTABLE ALLOCATIONS. In the previous two sections we introduced the notions of a coalitional preference and of an attainable allocation. Here both are combined to install a model-consistent comparison-rule - familiar to the blocking or improve upon rule - allowing the detection of allocations which are unacceptable redistributions with respect to the prevailing preferences and the given first endowments of the coalitions.

The definition is explained at first within the deterministic model to apply subsequently the work rule (W1).

Assume that $\mathcal{E} := E \times e : (\mathbb{N}, R_o, d) \to P_{mo} \times \mathbb{R}^n_+$ is a deterministic market with pure competition. Let $A \in R_o, A \neq \emptyset$, and let f,g be deterministic allocations. Then f is called A-*unacceptable by* g with respect to $\mathcal{E}$ provided that (a) $\lim_{n\to\infty}(1/n)\sum_{i=1,i\in A}^{n} g(i) \stackrel{\bullet}{\ll} \lim_{n\to\infty}(1/n)\sum_{i=1,i\in A}^{n} e(i)$, and (b) $g \succ_A f$ are fulfilled Here $x \stackrel{\bullet}{\ll} y$ iff $x_i < y_i > 0$ resp. $x_i = y_i = 0$ for $i=1,\ldots,n$.
We say that f is A-*unacceptable* with respect to $\mathcal{E}$ iff there exists a g such that f is A-unacceptable by g with respect to $\mathcal{E}$. Lastly f is *unacceptable* iff there exists a non void coalition A such that f is A-unacceptable. The above definitions are applied analogously with respect to the deterministic observed model.

Let us explain the meaning of the decision rules introduced above. Assume that the coalition A has knowledge about the allocation g. Due to the condition (a) the agents of A recognize that the alternative redistribution h, with $h|_{\mathbb{N}\setminus A} \equiv e$, $h|_A \equiv g$, can be reached without any help from other coalitions disjoint to A. Due to the condition (b) the offer to accept the alternative first endowment of commodities f has to be rejected by A. In analogy to the two previous sections the "A-unacceptable"- and the "unacceptable"-rules are constructed as collections of the above explained "A-unacceptable by"-rule.

At next we deduce that the introduced definitions are technical reformulations of the well known "blocking" resp. "improve upon" notions so far as finitely many agents are considered: If f is A-unacceptable by g then there is an allocation g' fulfilling (a') $\lim_{n\to\infty}(1/n)\sum_{i=1,i\in A}^{n} g'(i) = \lim_{n\to\infty}(1/n)\sum_{i=1,i\in A}^{n} e(i)$ and (b) $g' \succ_A f$. This is a consequence of the definition of $\succ_A$, i.e. g' can be obtained by adding g and $(\varepsilon_1\chi(A),\ldots,\varepsilon_n\chi(A))$ where $\chi(A)$ denotes the characteristic function of A and where $\varepsilon_1,\ldots,\varepsilon_n$ are suitably chosen positive numbers. Conversely, if g' fulfilling (a') and (b) is given then we obtain in the same manner a suitable g" fulfilling (a) and (b). Hence there is actually no difference between the "A-unacceptable" and the "blocking by A" or "improving by A" rule so far as finitely many agents are considered.

It remains to investigate the model-consistency of the introduced notions: Assume that f is A-unacceptable by g with respect to $\mathcal{E}$ and that $g\varepsilon - \succ_A f$ where $\varepsilon>0$ is known as well as f,g,A, and $\mathcal{E}$. Then we can *calculate* a fixed, finite degree of approximation guaranteeing that every finer measurement will recognize f to be A-unacceptable by g concerning $\mathcal{E}$. Hence it follows that in the cases "f is A-unacceptable by g", "f is A-unacceptable", and "f is unacceptable" there *exists* a fixed, finite precision guaranteeing a verification by approximation. We conclude that the introduced, deterministic definitions are consistent with the principles of our modelling, in particular with (A4). Therefore it remains to apply the work rule (W1) to obtain the final probabilistic definition.

<u>DEFINITION III.4.</u> Let $\bar{\mathcal{E}}:\bar{E}\times\bar{e}:(\mathrm{Stone}(R_o),\bar{\mathcal{R}}_o,\bar{d}) \to P_{mo}\times\mathbb{R}_+^n$ an observed market with pure competition. Let $\bar{f},\bar{g}\in C(\mathrm{Stone}(R_o),\bar{d},\mathbb{R}_+^n)$ are observed allocations and let $\bar{A}\in CO(\mathrm{Stone}(R_o))$ with $\bar{A}\neq\emptyset$. Then $\bar{f}$ is called $\bar{A}$-unacceptable by $\bar{g}$ with respect to $\bar{\mathcal{E}}$ provided that (A) $\int_{\bar{A}}\bar{g}d\bar{d} \,\dot{\ll}\, \int_{\bar{A}}\bar{e}d\bar{d}$ and (B) $\bar{g} \succ_{\bar{A}} \bar{f}$ are fulfilled. We denoted $x \,\dot{\ll}\, y$ iff $x_i<y_i>0$ resp. $x_i=y_i=0$ holds for i=1,...,n. We call $\bar{f}$ to be $\bar{A}$-unacceptable with respect to $\bar{\mathcal{E}}$ iff there exists a $\bar{g}$ such that $\bar{f}$ is $\bar{A}$-unacceptable by $\bar{g}$ with respect to $\bar{\mathcal{E}}$. Lastly $\bar{f}$ is unacceptable iff there exists a non void $\bar{A}\in CO(\mathrm{Stone}(R_o))$ such that $\bar{f}$ is $\bar{A}$-unacceptable. The restrictions onto $\mathbb{N}$ yield the related

notions concerning markets with pure competition.

The related definition concerning the elementarily represented model is obtained by replacing the notations in definition III.4 in the same manner as this was demonstrated in section III.5.4.

5.4 THE CORE. We are prepared now to introduce the final definition of this paragraph.

DEFINITION III.5. Assume that $\mathcal{E}$ is a (deterministic, resp. represented) market with pure competition. The Core $\mathcal{C}(\mathcal{E})$ of $\mathcal{E}$ consists of all attainable (deterministic, resp. represented), non unacceptable allocations. In the observed cases the definition holds analogously.

Although the above definition is verbally formulated we want to interpret its economic meaning: Assume that the allocation f is an element of the Core $\mathcal{C}(\mathcal{E})$ of $\mathcal{E}$. Then f is a possible redistribution of the given first endowment for f is attainable. Since f is not unacceptable there is no coalition which recognizes a reason to reject such an offered redistribution. This doesn't imply that any coalition recognizes an advantage by accepting the offered redistribution f. Hence the notion of the Core is an indirect tool characterizing those attainable allocations which remain after subtracting all unacceptable, attainable allocations.

The definition introduced above of the Core coincides with the classical definition provided that only finitely many agents are considered. Indeed our definition is based on the notions of attainable allocations and unacceptable allocations. These coincide with the classical definitions in the finite case.

It remains to discuss the model-consistency of the definition III.5. For that sake we consider again the case of finitely many agents: If $f \in \mathcal{C}(\mathcal{E})$ then there exists no finite precision allowing us to recognize f as an element of the Core. This is a direct consequence of the indirect way in which $\mathcal{C}(\mathcal{E})$ is defined. As the Core is not an inherently, numerically stable tool within the finite situation we have to accept that fact within the case of pure competition as well. Indeed the Core is a historically grounded, qualitative instrument for characterizing possible redistributions of the considered first endowment. Our introducing of the Core is mainly caused by the wish to follow and reflect the historically developed thread of the theory.

Instead of the Core one can equivalently consider the *set* U($\mathcal{E}$) *of all attainable, unacceptable allocations*. The construction of this set is consistent with the inherent principles of our model.

5.5 CONCLUSION. In this part of the paper we introduced the classical definitions of an attainable, a preferred, and an unacceptable (or blocked) allocation in consistency with the modelling principles of the developed market with pure competition. The conformity with the classical definitions concerning markets with finitely many agents was demonstrated. The notion of the Core, equally based on the above definitions as in the case of a finite market, was introduced and its lack of numerical stability was analyzed. As a consequence we suggested the model-consistent, equivalent definition of the set of unacceptable, attainable allocations.

The main reason for the presented, static analysis of the possible redistributions of the considered first endowment was our intention to probe and demonstrate the aptitude of the developed market model for studies in problems of the classical theory of pure competition. It was not our goal to discuss and analyze the classical framework in itself with respect to its ability in solving microeconomic questions.

6. WALRAS ALLOCATIONS

As well as in the previous part III.5 we restrict ourselves here onto a model-consistent installation or translation of the classical definition in such a way that the reduction on markets with finitely many agents delivers the historically based notions.

6.1 PRICE SYSTEMS. In the sixth part of the work we want to study within a static analysis the possible redistributions of the first endowment of a market with the help of an additional economic instrument namely an imposed, fixed system of prices for the considered commodities prescribing the pairwise, quantitative relations at which exchanges at best are admitted. Before a more complete discussion and interpretation of that additional tool of our investigation, the necessary notations shall be introduced:

DEFINITION III.6. We denote $\mathbb{R}^n_{++} := \{x \in \mathbb{R}^n \mid x \gg 0\} = \{x = (x_1, \ldots, x_n) \mid x_i > 0$ for $1 \leq i \leq n\}$. If $p \in \mathbb{R}^n_{++}$ then the ray $\mathbb{p} := \{\lambda p \mid \lambda \in \mathbb{R}, \lambda > 0\}$ is called a price system. Every $p' \in \mathbb{p}$ is called a price vector with respect to $\mathbb{p}$ (or a representative of the price system $\mathbb{p}$).

Assume that $p := (p_1, \ldots, p_n) \in \mathbb{R}^n_{++}$ is a given price vector. Then p installs a system of relations as follows: We imagine that one unit of the i-th commodity can be exchanged at best to (p_i/p_j) units of the j-th commodity. These relations derived from the price vector p will allow us to install restrictions on the set of alternative allocations which we want to compare with the given first endowment of commodities within a static analysis. One recognizes that all price vectors derived from a given price system $\mathbb{p}$ are equivalent with respect to the relations described above.

The notions introduced above of a price system and a price vector are interpreted by us as an instrument for a *static* analysis. We emphasize that the notions used here should not be misunderstood as prices arising in markets of reality. Prices in markets of reality are a result of a *dynamic* process based on the dynamic behaviour and acting of the members of these markets. Only in very special situations may our tool of a static analysis be comparable to prices of a market in reality.

When we keep prices of a market in reality distinct from the notion of a price vector used here as an instrument for a static analysis then we abolish by no means our conviction that elements of $\mathbb{R}^n$ remain idealized objects deriving their relevance from an, at least principally possible, approximation framework based on calculations in reality.

No problem arises in approximating a price vector p with respect to the Euclidean distance by e.g. rational vectors. A quantitative approximation of a price system $\mathbb{p}$ would require more technical notations. We are interested within our theoretical, static analysis in the possibility of an approximation only. Therefore we will work subsequently with price vectors and with the usual Euclidean metric to explain and discuss the model consistency with respect to involved price vectors.

6.2 <u>p-ATTAINABLE ALLOCATIONS</u>. Here the p-dependent restrictions which we want to impose on the set of investigated allocations of this sixth part are elaborated.

We start by considering the deterministic model. Let $\mathcal{E}:=E\times e:(\mathbb{N},R_o,d)\to P_{mo}\times\mathbb{R}^n_+$ a deterministic market with pure competition and let $p\in\mathbb{R}^n_{++}$ be a given price vector. Let f be a deterministic allocation and let $\varepsilon\in\mathbb{R},\varepsilon>0$. Then f is called ε-p-*attainable* with respect to $\mathcal{E}$ iff (B) $(\sum_{i=1}^n p_i f_i(j))+\varepsilon<\sum_{i=1}^n p_i e_i(j)$ is fulfilled for every agent $j\in\mathbb{N}$. We call f *positive*-p-*attainable* iff there exists an $\varepsilon>0$ such that f is ε-p-attainable. Lastly f is p-*attainable* iff it can be approximated concerning the sup. norm by allocations which are positive-p-attainable with respect to $\mathcal{E}$. In the case of the observed model we apply the above definitions analogously, i.e. we require in this case that the above inequality (B) is fulfilled for every $x\in\text{Stone}(R_o)$.

<u>REMARK III.7.</u> Let $\mathcal{E}$ be a deterministic market with pure competition and let p be a price vector. A deterministic allocation f is positive-p-attainable with respect to $\mathcal{E}$ iff the observed, deterministic allocation $\bar{f}$ is positive-p-attainable with respect to $\bar{\mathcal{E}}$. Moreover, f is p-

attainable iff $\bar{f}$ is p-attainable.

The proof of the above remark III.7 is obvious and analogous to the proof of prop. III.1. Moreover, the above remark III.7 is a hint at the model consistency of the introduced definitions. Indeed if f is positive-p-attainable then there exists a fixed, finite precision such that every finer approximation of p, f, and e verifies that property. The application of the work rule (W1) yields the related definitions for the probabilistic model:

<u>DEFINITION III.8.</u> Let $\bar{\mathcal{E}}:=\bar{E}\times\bar{e}:(\mathrm{Stone}(R_o), \bar{\mathcal{R}}_o, \bar{d}) \to P_{mo} \times \mathbb{R}^n_+$ an observed market with pure competition. Let $\bar{f}$ be an observed allocation and let p be a price vector. (i) Let $\varepsilon \in \mathbb{R}, \varepsilon > 0$. Then $\bar{f}$ is called ε-p-attainable with respect to $\bar{\mathcal{E}}$ iff $(\bar{B})$ $(\sum_{i=1}^n p_i \bar{f}_i(x)) + \varepsilon < \sum_{i=1}^n p_i \bar{e}_i(x)$ is fulfilled for $x \in \mathrm{Stone}(R_o)$ except on a set A with $\bar{d}(\mathrm{cl}(A)) = 0$. Here A may depend on $\bar{f}$, $\bar{\mathcal{E}}$, p, and ε. (ii) We call $\bar{f}$ positive-p-attainable with respect to $\bar{\mathcal{E}}$ iff there exists an $\varepsilon > 0$ such that $\bar{f}$ is ε-p-attainable with respect to $\bar{\mathcal{E}}$. (iii) $\bar{f}$ is called p-attainable iff it can be approximated concerning the ess. sup. norm by allocations which are positive-p-attainable with respect to $\bar{\mathcal{E}}$. The restriction onto $\mathbb{N} \subset \mathrm{Stone}(R_o)$ yields the related definitions for a market with pure competition.

Naturally the remark III.7 holds in the probabilistic situation in strict analogy to the deterministic one. The related definitions for the elementarily represented model are obtained from definition III.8 by replacing the notations as it was demonstrated at the end of section III.5.1 resp. III.5.2.

If we restrict ourselves onto the economic relevant case - i.e. a market with pure competition and a *positive first endowment* - then the above considered set of positive-p-attainable allocations is *not empty*. A study of the arbitrary case requires a more technical reformulation of the above definition.

It remains to compare the above introduced notions with the classical definitions which are well known from markets with finitely many agents. If we consider a "market" with one single agent then we recognize immediately that the set of p-attainable allocations coincides with the *budget set* of the agent concerning the price vector p. Analogously, if finitely many agents are considered then the set of all p-attainable allocations turns out to consist of all those allocations which satisfy the budget restrictions of the agents. Hence we conclude that our notions are a reformulation of the classical ones. Due to that reformulation the set of p-attainable allocations is the completion of the set of all positive-p-allocations. Since these are constructed strictly following the grounding axiom (A4), resp. (A4*), we consider the above definition to be model consistent.

6.3 <u>p-UNACCEPTABLE ALLOCATIONS.</u> The p-attainable allocations subject to the agents' budget restrictions are compared now with the help of the coalitional-preference approach developed in section III.5.2.

<u>DEFINITION III.9.</u> Let $\mathcal{E}$ be a market with pure competition, and let p be a price vector. A p-attainable allocation f is called p-unacceptable iff there exists a positive p-attainable allocation g and a non void coalition A with $g \succ_A f$. In the case of an observed (represented, deterministic) market the notions are applied analogously.

The above definition is a model-consistent combination of the definition III.8 and the coalitional-preference approach developed in section III.5.2. Indeed, assume that f is p-unacceptable. Then there is an allocation g_o, a non-void coalition A_o, and an $\varepsilon>0$ such that g_o is ε-p-attainable and $g_o \varepsilon\text{-} \succ_{A_o} f$. Hence, a sufficiently precise approximation of p, g_o, f, and the status of pure competition enables us to recognize that f is p-unacceptable.

Note that in analogy to section III.5.3 the following can be deduced: Let f be p-attainable. If there is a positive-p-attainable allocation g and a coalition A with $g \succ_A f$ then there is a p-attainable allocation g' fulfilling the budget-equality restriction such that $g' \succ_A f$.

This is a consequence of the monotonicity of market mappings. On the other hand, a p-attainable allocation g" fulfilling the budget-equality restriction and fulfilling $g'' \succ_A f$ yields a positive-p-attainable allocation g with $g \succ_A f$.

We want to compare our notion with the classical ones. For that sake we define: A p-attainable allocation f is called a p-*demand allocation* if f is not p-unacceptable. Due to the paragraph above we recognize that our p-demand allocation coincides with a classical demand function and vice versa provided that finitely many agents are considered. Consequently we consider our approach as a model-consistent one which coincides with the classical framework within the finite case.

REMARK III.10. Let $\overline{\mathcal{E}}$ be a deterministic, observed market with pure competition and a positive first endowment. Let p be a price vector. Let $\bar{f}$ be a deterministic, observed p-demand allocation with respect to $\overline{\mathcal{E}}$. Then $\bar{f}(x)$ is an optimal element in the budget set of every $x \in \mathrm{Stone}(R_o)$ with respect to $\bar{E}(x)$.

At the end of this section we want to add two comments: (i) Our definition of a p-demand allocation is an indirect one similar to our definition of the Core. This is a consequence of the fact that the notion of a maximum or an optimum is a purely mathematical one. The definition of an optimum is in general *not* a stable tool of the numerical analysis. Only if there are *additional* nice properties of the considered problem then a numerical stable algorithm exists for an approximation of an optimum. (ii) In the case of finitely many agents a p-demand allocation is interpreted as an *independent* choice made by the agents within the optimal commodity bundles of their p-budget sets. Therefore one may suspect this interpretation to be abolished within our framework for we are considering continuous functions on $\mathrm{Stone}(R_o)$ only. However, remember that our analysis started in chapter I with an *arbitrary* set of agents and an *arbitrary*, bounded first endowment of commodities. Then we deduced carefully the transformation of these markets into the replica model. Hence we conclude that the picture of an independent choice is *not* necessarily abolished. Indeed the restriction onto the replica model is essentially a consequence of the fact that P as well as a bounded region of $\mathbb{R}^n_+$ are compact. A sequence in a compact set always possesses accumulation points. Therefore an enumeration consistent with our modelling is possible.

6.4 p-WALRAS ALLOCATIONS. Within our static analysis we want to study here allocations which are attainable as well as p-attainable.

DEFINITION III.11. Let $\mathcal{E}$ be a market with pure competition and let p be a price vector. An attainable, p-attainable, and not p-unacceptable allocation with respect to $\mathcal{E}$ is called a p-Walras allocation with respect to $\mathcal{E}$. An allocation f is a Walras allocation with respect to $\mathcal{E}$ iff there is a price vector p such that f is a p-Walras allocation concerning $\mathcal{E}$. We denote $W(\mathcal{E})$ the set of all Walras allocations of $\mathcal{E}$. The definition is applied analogously for markets which are observed or represented.

By the above definition we recognize that a p-Walras allocation is an *attainable* p-*demand allocation*. Hence the notion of a p-Walras or a Walras allocation is an instrument of a static, microeconomic analysis similarly indirect as the notion of the Core is. Indeed if f is a p-Walras allocation then we are not capable of verifying that property by a sufficiently precise approximation of f, p, and $\mathcal{E}$.

Our notion of a p-Walras resp. of a Walras allocation coincides with the classical one provided that finitely many agents are considered and that a positive first endowment prevails. This follows because the definition of a p-Walras allocation is a combination of the notions of an attainable, a p-attainable, and a p-unacceptable allocation.

At the end the reader should allow us to mention an elementary warning: Provided a price vector p allowing a p-Walras allocation f is considered then we can not conclude that every further p-demand allocation is attainable as well. This may be false even in the case of two agents if there are not additional assumptions about the agent's preferences. Hence a p-Walras allocation should not be interpreted as an independent choice of commodity bundles by the agents as it was accepted by us with respect to the p-demand allocations. A p-Walras allocation is a *static*, microeconomic characterization of allocations. In special cases it may be interpreted as the result of a *dynamic* process creating prices similarly as in real markets.

6.5 CONCLUSION. Within this sixth part of the paper we probed the possibility to translate within the framework of our model the classical notion of a Walras equilibrium.

A Walras equilibrium was interpreted by us as a possible outcome from a *static* equilibrium analysis of the considered market model with respect to a *static* instrument of investigation, namely an imposed, fixed price system prescribing to the agents the binding exchange rates. A competitive equilibrium in a *real* market - i.e. the actual commodity allocation and the prevailing price system at a fixed time point - was seen by us as the result of the previous *dynamical* acting and bargaining of all of the members of the considered market in reality. Hence a Walras equilibrium was deemed by us to be the notice or the snapshot of an evolving equilibrium time path at a fixed instant similarly e.g. as in macroeconomics a static analysis may be understood as the restriction of a superpositioned, dynamical study.

Emanating from our above depicted, principal view of the nature of Walras' equilibria we translated the classical notions in analogy to the technics used in part III.5 by combining the inherent principles of our modelling with the requirements prescribed by the finite case:

A price system $\mathbb{P}$ given by a price vector $p \in \mathbb{R}^n_{++}$ was introduced as an instrument for a static, microeconomic analysis where its idealized character was not concealed. Provided that a commodity allocation fulfilled the budget restrictions of every agent in an observable, model consistent manner then it was called a positive-p-attainable allocation. The completion of the set of all positive-p-attainable allocations yielded the p-attainable allocations in conformity with the classical notion. A p-attainable allocation was declared to be p-unacceptable provided that a group of agents could recognize it in a model consistent manner as not preferable over their first endowment. A p-attainable, but not p-unacceptable allocation was called a p-demand allocation in correlation with the classical definition for the finite case. As well as in part III.5.4 the nature of a mathematical optimum caused the indirect way of the definition. A p-demand allocation was still interpreted as an independent, optimal choice of commodity bundles by the agents subject to their budget restrictions. A p-Walras equilibrium was an attainable p-demand allocation. In the finite situation we obtained in this way the familiar notion. In general a Walras equilibrium could not be interpreted as an independent, optimal choice of commodity bundles subject to a prevailing, suitable price system. This fact yielded *no* contradiction to our interpretation of Walras' equilibria as a *static* tool.

7. CORE VERSUS WALRAS ALLOCATIONS

The adapting of the classical definitions within our market model could be installed in the previous part 5 and part 6. Hence we probe now whether the classical conjecture can be verified.

PROPOSITION III.12. Let $\mathcal{E} := E \times e : (\mathbb{N}, R_o, d) \to P_{mo} \times \mathbb{R}^n_+$ a market with pure competition and a positive first endowment. Then $W(\mathcal{E}) = \mathcal{C}(\mathcal{E})$.

P r o o f: We prove that $W(\mathcal{E}) \subset \mathcal{C}(\mathcal{E})$: To avoid unnecessary notations we restrict ourselves onto the deterministic case. Assume that there is a $f \in W(\mathcal{E})$ but $f \notin \mathcal{C}(\mathcal{E})$. Then there is a non void coalition $A \in R_o$ and an allocation g with (i) $g \succ_A f$ and (ii) $\lim_{n\to\infty}(1/n)\sum_{i=1,i\in A}^{n} g(i) = \lim_{n\to\infty}(1/n)\sum_{i=1,i\in A}^{n} e(i)$. By (i) and the remark III.10 we obtain $p\cdot e(i) < p\cdot g(i)$ for $i \in A$ where p is the price vector related to f. Inserting this inequality into (ii) yields a contradiction. Hence $W(\mathcal{E}) \subset \mathcal{C}(\mathcal{E})$.

Due to its shortness we carried out explicitely the first part of the proof. The converse direction turns out to be an application of the general result of T.E.ARMSTRONG and M.K.RICHTER [2]. Since the explicit presentation of the second part of the proof requires, up to a certain stage, a recapitulation of their approach we reserve this technical step for the appendix.

Naturally the result stated above holds as well for observed or represented markets.

Probably the reader is waiting now for the presentation of the existence result with respect to $W(\mathcal{E})$. Two reasons occasion us to drop such an attempt. At first we could invite the reader to try an application of the general result of T.E.ARMSTRONG and M.K.RICHTER [3] as well as in the above proof. However the main reason is the following, second one: We have not rigorously studied within this treatment the space of preferences P concerning its relation to microeconomic measurements. We accepted P only as a mathematical tool which is sufficient with respect to the situation studied here, namely the modelling of the status of pure competition. The existence result however is more involved by properties of the considered preferences. Hence before a model consistent study of the existence theorem is made, a more rigorous study of the concept of the space of preferences seems to us to be necessary. In contrast to the dependence of the existence result on the properties of preferences, the equivalence result is a characteristic outcome of the status of pure competition. Therefore its presentation here is deemed by us to be justified.

By the result presented above we have demonstrated that the model developed here of the status of pure competition is suitable to answer classical questions. Its precise interpretability and concreteness is no disavantage with respect to the application of abstract concepts developed within this field.

APPENDIX

This appendix treats of the following three main topics: At first a short introduction into the theory of Boolean algebras, Stone spaces, and related material form measure theory as well as an introduction into the generalized Riemann-Darboux integration theory is presented. References concerning the proofs omitted here are indicated. Secondly we collect some technical proofs which were refered only within the first three chapters. Lastly we gathered a unified rendering of the elementary representation of the model.

A1 BOOLEAN ALGEBRAS AND STONE SPACES

In this part we put some facts together concerning Boolean algebras, Stone spaces and measure theory which may be helpful to non-experts. Proofs concerning Boolean algebras and Stone spaces can be verified in the introductory book of P.R.HALMOS [A3] or the monographs of R.SIKORSKI [35] and D.A.VLADIMIROV [36]. For the measure theoretic tools we refer the reader to P.R.HALMOS [11] or Z.SEMADENI [33].

A.1.1 BOOLEAN ALGEBRAS. A rigorous, axiomatic, and general definition of a Boolean algebra is too technical and, in our opinion, not absolutely necessary for a first comprehension of our paper, especially of the first chapter. Therefore we restrict ourselves here onto a special kind of Boolean algebras, also called fields, aiming mainly at non-experts.

Let N be a set and let $\mathbb{P}(N)$ be the power set of N; i.e. $\mathbb{P}(N)$ is the class of all subsets of N. Moreover let $\mathfrak{A}$ be a subset of $\mathbb{P}(N)$ with $N \in \mathfrak{A}$

and $\emptyset\in\mathfrak{A}$; i.e. $\mathfrak{A}$ is a collection of subsets of N including N and the empty set. We assume that $\mathfrak{A}$ is closed with respect to finite combinations of the union, intersection, and complement operation; i.e. if $A_1,\ldots,A_n$ are elements of $\mathfrak{A}$ then we require that always $B_1 \oplus_1 B_2 \oplus_2 \cdots \oplus_{n-1} B_n \in \mathfrak{A}$ where $B_i=A_i$ or $B_i=N\setminus A_i=:-A_i$ and where $\oplus_i \equiv \cup$ or $\oplus_i \equiv \cap$. Then $\mathfrak{A}$ is called a *Boolean* (set) *algebra* (or also a field).

EXAMPLE A.1.1. Let N be a set. (a) If $\mathfrak{A}:=\{\emptyset,N\}$ then $\mathfrak{A}$ is a Boolean algebra. (b) The power set $\mathbb{P}(N)$ itself is a Boolean algebra. (c) Let $\mathfrak{A}:=\{A\subset N \mid A$ is finite or the complement $-A$ is finite$\}$. Then $\mathfrak{A}$ is a Boolean algebra.

Let N,N' be two sets and let $\mathfrak{A}\subset\mathbb{P}(N)$, $\mathfrak{A}'\subset\mathbb{P}(N')$, be two Boolean algebras. A mapping $h:\mathfrak{A}\to\mathfrak{A}'$ is called a *Boolean homomorphism* if h preserves the union, intersection, and complement operation; i.e. if

(a) $h(A\cup B)=h(A)\cup h(B)$

(a') $h(A\cap B)=h(A)\cap h(B)$

(b) $h(-A)=-h(A)$

holds for every $A,B\in\mathfrak{A}$.

We mention some properties of Boolean homomorphisms: By de Morgan's formulas (a), (b) imply (a') and (a'), (b) imply (a). Moreover $h(\emptyset)=\emptyset$ and $h(N)=N'$. If $A\subset B\subset N$ then $h(A)\subset h(B)$ and $h(A\setminus B)=h(A)\setminus h(B)$.

If the Boolean homomorphism $h:\mathfrak{A}\to\mathfrak{A}'$ is bijective then h is called a *Boolean isomorphism*. In this case h^{-1} is a Boolean isomorphism from $\mathfrak{A}'$ onto $\mathfrak{A}$. We mention that h is a Boolean isomorphism provided that $A\subset B$ iff $h(A)\subset h(B)$.

EXAMPLE A.1.2. (a) We consider the set $N=\mathbb{N}$ of natural numbers and the one-point-set $\Pi\in\mathbb{R}$. Let $\mathfrak{A}:=\{A\subset N \mid A$ is finite or $-A:=\mathbb{N}\setminus A$ is finite$\}$ $\subset\mathbb{P}(\mathbb{N})$ and let $\mathfrak{A}':=\{\emptyset,\Pi\}=\mathbb{P}(\Pi)$. Then $h:\mathfrak{A}\to\mathfrak{A}'$ defined by $h(A):=\emptyset$, if A is finite, and $h(A):=\Pi$, if $-A$ is finite, is a Boolean homomorphism. (b) Let $\mathfrak{A}'$ be defined as above. Let $\mathfrak{A}:=\{\emptyset,\mathbb{N}\}\subset\mathbb{P}(\mathbb{N})$. Then $h:\mathfrak{A}\to\mathfrak{A}'$ defined by $h(\emptyset)=\emptyset$ and $h(\mathbb{N})=\Pi$ is a Boolean isomorphism. Naturally $\mathbb{N}$ is not bijective to the one-point-set Π. (c) Due to G.Cantor there exists a bijection $\varphi:\mathbb{N}\xrightarrow{\cong}\mathbb{Q}$ between the natural numbers and the rational ones. Obviously φ induces then a Boolean isomorphism between $\mathbb{P}(\mathbb{N})$ and $\mathbb{P}(\mathbb{Q})$.

A.1.2 STONE SPACES. A *filter* in the Boolean algebra $\mathcal{O}\subset\mathbb{P}(N)$ is a non-empty subset $\mathcal{F}\subset\mathcal{O}\setminus\{\emptyset\}$ fulfilling: (a) if $A,B\in\mathcal{F}$ then $A\cap B\in\mathcal{F}$ and (b) if $A\in\mathcal{F}$, $B\in\mathcal{O}$, $A\subset B$ then $B\in\mathcal{F}$. A filter $\mathcal{F}$ in $\mathcal{O}$ is called an *ultra-filter* provided that $A\in\mathcal{F}$ or $-A\in\mathcal{F}$ holds for every $A\in\mathcal{O}$. The case $A\in\mathcal{F}$ and $-A\in\mathcal{F}$ never occurs for $A\cap -A=\emptyset\notin\mathcal{F}$.

EXAMPLE A.1.3. We consider again the example A.1.2 (a) where $\mathcal{O}$ was the Boolean algebra of finite and cofinite subsets of the natural numbers. Let $n\in\mathbb{N}$. Then $\mathcal{F}_n:=\{A\in\mathcal{O}\mid n\in A\}$ is a filter and an ultrafilter. Such an ultrafilter determined by a point of the underlying set is called a *fixed ultrafilter*. Let $\mathcal{F}_\infty:=\{A\in\mathcal{O}\mid A \text{ is not finite}\}$. Then $\mathcal{F}_\infty$ is an ultrafilter in $\mathcal{O}$ which is not determined by a point $n\in\mathbb{N}$. Such an ultrafilter is called a *free ultrafilter*. Let us consider the set $\gamma\mathbb{N}:=\mathbb{N}\cup\{\infty\}$; i.e. $\gamma\mathbb{N}$ consists of $\mathbb{N}$ and the additional, symbolic point ∞. Let $\mathcal{B}\subset\mathbb{P}(\gamma\mathbb{N})$ be defined as follows: $\mathcal{B}:=\{A\mid$ either $A\subset\mathbb{N}$ and A is finite or $\infty\in A$ and $-\mathbb{N}\cap A$ is finite$\}$. Then $\mathcal{B}$ is a Boolean algebra and $\mathcal{B}$ is Boolean isomorphic to $\mathcal{O}$. Moreover every ultrafilter in $\mathcal{B}$ is fixed: Indeed an ultrafilter in $\mathcal{B}$ is determined either by a point $n\in\mathbb{N}$ or by the symbolic point ∞. Hence we have represented $\mathcal{O}$ by $\mathcal{B}$ such that $\mathcal{B}$ possesses fixed ultrafilters only. We describe subsequently the analogous representation for the general situation.

A Boolean algebra $\mathcal{O}\subset\mathbb{P}(N)$ is considered where N is a set. We define $\text{Stone}(\mathcal{O}):=\{x\mid x \text{ is an ultrafilter in } \mathcal{O}\}$. The set $\text{Stone}(\mathcal{O})$ is called the *Stone space* of $\mathcal{O}$. If $A\in\mathcal{O}$ then we define $\bar{A}:=\{x\in\text{Stone}(\mathcal{O})\mid A\in x\}$. Then $\bar{A}$ is a subset of $\text{Stone}(\mathcal{O})$ and therefore an element of the power set $\mathbb{P}(\text{Stone}(\mathcal{O}))$ of the set $\text{Stone}(\mathcal{O})$. Let $\bar{\mathcal{O}}:=\{\bar{A}\mid A\in\mathcal{O}\}$. Then $\bar{\mathcal{O}}\subset\mathbb{P}(\text{Stone}(\mathcal{O}))$ is a Boolean algebra. Moreover $h:\mathcal{O}\to\bar{\mathcal{O}}$ defined by $A\mapsto\bar{A}$ is a *Boolean isomorphism. Every* ultrafilter in $\bar{\mathcal{O}}$ is a *fixed ultrafilter:* Indeed, if $\bar{\mathcal{F}}$ is an ultrafilter in $\bar{\mathcal{O}}$ then we obtain by

$h^{-1}\{\bar{F} \mid \bar{F} \in \bar{\mathfrak{F}}\}$ an ultrafilter $x \in \mathfrak{A}$. Then x is a point in Stone($\mathfrak{A}$) and x determines $\bar{\mathfrak{F}}$; i.e. $\bar{\mathfrak{F}} = \{\bar{A} \in \bar{\mathfrak{A}} \mid x \in \bar{A}\}$.

In a next step we want to introduce a topology on the Stone space Stone($\mathfrak{A}$). Remember that the topology of $\mathbb{R}$ consists of the system of all open subsets of $\mathbb{R}$. This collection possesses the following two properties: (a) The union of an arbitrary class of open sets is itself an open set; especially the whole set - i.e. $\mathbb{R}$ - is open. (b) The intersection of finitely many open sets is open; in particular the empty set is open. The general case of a topological space is characterized in the same manner: Let N be a set and let $\mathcal{T}\mathcal{O} \subset \mathbb{P}(N)$ be a system of subsets of N fulfilling the above two conditions (a) and (b). Then the pair $(N, \mathcal{T}\mathcal{O})$ is called a *topological space* and $\mathcal{T}\mathcal{O}$ is its *topology*.

In most of the cases an explicit definition of $\mathcal{T}\mathcal{O}$ is too complicated and not necessary. Instead of that one introduces usually a special class $\mathcal{O} \subset \mathbb{P}(N)$ requirering that $\mathcal{T}\mathcal{O}$ is the smallest system in $\mathbb{P}(N)$ subject to (a), (b), and $\mathcal{O} \subset \mathcal{T}\mathcal{O}$; e.g. with respect to $\mathbb{R}$ the knowledge of open intervals is in general sufficient.

Analogously as above we introduce a topology $\mathcal{T}\mathcal{O}$ for the Stone space: Let N be a set and let $\mathfrak{A} \subset \mathbb{P}(N)$ be a Boolean algebra. If $A \in \mathfrak{A}$ then we define $\bar{A} \subset$ Stone($\mathfrak{A}$) to be *open*.. Let $\mathcal{O} := \{\bar{A} \subset \text{Stone}(\mathfrak{A}) \mid A \in \mathfrak{A}\}$. Then $\mathcal{T}\mathcal{O}$ is the smallest subclass of $\mathbb{P}$(Stone($\mathfrak{A}$)) fulfilling the conditions (a), (b) and $\mathcal{O} \subset \mathcal{T}\mathcal{O}$. Hence, *(Stone($\mathfrak{A}$), $\mathcal{T}\mathcal{O}$) is a topological space*. Observe that the elements of $\mathcal{O}$ are as well closed sets because $-\bar{A} = \overline{(-A)}$ is open. Hence, if $A \in \mathfrak{A}$ then $\bar{A}$ is a closed *and* open - abbreviated: *clopen* - subset of Stone($\mathfrak{A}$). Studying the Stone space tool more precisely one can prove that $\mathcal{O} := \{\bar{A} \subset \text{Stone}(\mathfrak{A}) \mid A \in \mathfrak{A}\}$ is actually the system of *all* clopen subsets of Stone($\mathfrak{A}$) concerning the topology $\mathcal{T}\mathcal{O}$. Therefore *a Boolean algebra $\mathfrak{A}$ is Boolean isomorphic to the field of all clopen subsets of its Stone space.*

EXAMPLE A.1.4 We investigate once more the Boolean algebra $\mathfrak{A}$ of finite and cofinite subsets of $\mathbb{N}$. In example A.1.3 we calculated the Stone space of $\mathfrak{A}$. We obtained Stone($\mathfrak{A}$)=$\mathbb{N}\cup\{\infty\}$; $\bar{A}=A$ if A is finite; $\bar{A}=A\cup\{\infty\}$ if A is cofinite. With respect to the Stone space topology every finite subset of $\mathbb{N}$ is clopen i.e. every $n\in\mathbb{N}$ is a clopen set in $\mathbb{N}\cup\{\infty\}$. The sets $A\cup\{\infty\}$, where A is cofinite, form a clopen neighbourhood basis system of the point ∞. Let $S:=\{n^{-1}\mid n\in\mathbb{N}\}\cup\{0\}$. Then g:Stone($\mathfrak{A}$)$\rightarrow S$ defined by $g(\infty)=0$ and $g(n)=n^{-1}$ for $n\in\mathbb{N}$ yields a bijection. We endow S with the topology induced by $\mathbb{R}$, i.e. $A\subset S$ is open iff there is an open set $A'\subset\mathbb{R}$ with $A'\cap S=A$. Then S is homeomorphic to Stone($\mathfrak{A}$). Hence S is a familiar picture and a representation of Stone($\mathfrak{A}$).

Lastly we mention an important property of Stone spaces: A topological space is by definition *compact* if every open covering of it contains a finite subcovering. One can prove that *the Stone space of a Boolean algebra is compact.*

We tried to introduce some facts concerning Boolean algebras and Stone spaces in an elementary manner. Since a Stone space is a kind of a completion it remains necessarily an abstract tool. Probably it still creates some aversion in many researchers in economics. But remind that the real numbers widely familiar today are an abstract completion tool as well. Indeed the names 'irrational number' or 'transcendent number' are an obvious hint at the complexity of $\mathbb{R}$.

A.1.3 MEASURE THEORETIC TOOLS. It remains to collect some facts concerning measure theory: Let N be a set and let $\mathfrak{A}\subset\mathbb{P}(N)$ a Boolean algebra. A mapping $\alpha:\mathfrak{A}\rightarrow[0,1]$ with $\alpha(\emptyset)=0$, $\alpha(N)=1$ is called a *finitely additive measure* on $\mathfrak{A}$ if $\alpha(A\cup B)=\alpha(A)+\alpha(B)$ holds always for $A,B\in\mathfrak{A}$, $A\cap B=\emptyset$. The mapping α is called a *measure* on $\mathfrak{A}$ if additionally $\alpha(A_o)=\sum_{i=1}^{\infty}\alpha(A_i)$ holds for every class $\{A_i\}_{i=o}^{\infty}\subset\mathfrak{A}$ with $A_i\cap A_j=\emptyset$ for $i,j\in\mathbb{N}$, $i\neq j$, and $A_o=\cup_{i=1}^{\infty}A_i$.

Let $\mathfrak{B}$ be the Boolean algebra of clopen subsets of Stone($\mathfrak{A}$). Then $\mathfrak{A}$ is Boolean isomorphic to $\mathfrak{B}$ by $A\mapsto\bar{A}$. A finitely additive measure on $\mathcal{A}$ creates a finitely additive measure β on $\mathcal{B}$ by $\alpha(A)=:\beta(\bar{A})$. Then β *is a measure on* $\mathcal{B}$: Indeed, since a clopen subset of Stone($\mathcal{A}$) is compact, there exists no system $\{\bar{A}_i\}_{i=0}^{\infty} \subset \mathcal{B}$ with $\bar{A}_i \cap \bar{A}_j = \emptyset$ for $i,j \in \mathbb{N}$, $i \neq j$, and $\bar{A}_0 = \cup_{i=1}^{\infty} \bar{A}_i$. Provided that such a system would exist then $\bar{A}_0$ would not fulfil the compactness condition mentioned above.

Since β is a measure on $\mathcal{B}$ there is a unique Radon measure $\bar{\beta}$ defined on the Borel-σ-field of Stone($\mathcal{A}$) and extending β (see P.R.HALMOS [11] § 13 Theorem A, § 54 Theorem D and Z.SEMADENI [33] 18.1.4 Exercise (c) for explicit proofs). Here a Radon measure is a regular, σ-additive measure.

EXAMPLE A.1.5 Let $\mathcal{A}$ be the Boolean algebra of finite and cofinite subsets of $\mathbb{N}$. Let $\alpha : \mathcal{A} \rightarrow [0,1]$ be defined by $\alpha(A)=0$, if A is finite, and $\alpha(A)=1$ otherwise. Then $\bar{\beta}$ is the Dirac measure centered at the point ∞ of Stone($\mathcal{A}$); i.e. $\bar{\beta}(E)=1$, if $\infty \in E$, and $\bar{\beta}(E)=0$, if $\infty \notin E$, where $E \subset$ Stone($\mathcal{A}$).

A2 THE RIEMANN-DARBOUX INTEGRATION

In this second part of the appendix we collect definitions and properties of the generalized Jordan content and the Riemann-Darboux integration theory. Although the definition of the general Riemann integral is quite similar to its classical counterpart there remain many worth mentioning properties concerning this tool. Often these properties are unfamiliar even in the classical case. For proofs we refer the reader to the articles [15], [16], [17], [18] of S.ROLEWICZ and the author.

A.2.1 THE GENERALIZED JORDAN CONTENT. Let K be a compact (Hausdorff-)space. $\mathcal{B}(K)$ denotes the Borel-σ-field on K, i.e. the smallest σ-field on K containing all compact subsets of K. Assume that μ is a non negative Radon measure on K; i.e. μ is a non negative, regular, σ-additive measure defined on $\mathcal{B}(K)$. Remind that μ is *regular* iff for every $A\in \mathcal{B}(K)$ and for every $\varepsilon>0$ there is a closed set $E(A,\varepsilon)$, an open set $F(A,\varepsilon)$ with $E(A,\varepsilon)\subset A\subset F(A,\varepsilon)$ and $\mu(F(A,\varepsilon)\setminus E(A,\varepsilon))<\varepsilon$. For convenience we shall assume μ to be *normalized* i.e. $\mu(K)=1$.

Let A be a subset of K. The *boundary* of A is $\partial A:=cl(A)\cap cl(K\setminus A)$ where cl(A) denotes the closure of A. If $\mu(\partial A)=0$ then A is called a *μ-continuity set* (or also a μ-Jordan measurable set resp. a μ-Jordan set). $\mathcal{B}(K,\mu)$ is the *class of all* μ-continuity subsets of K. $\mathcal{B}(K,\mu)$ is a field but in general not a σ-field.

We denote int(A) the interior of A. If A=cl int(A) then A is *regular closed.* A subset A of K is a *μ-Jordan zero set* iff $\mu(cl(A))=0$. Then $Z(K,\mu)$ denotes the *class of all* μ-Jordan zero subsets of K. Obviously $Z(K,\mu)\subset \mathcal{B}(K,\mu)$. We denote $A\Delta B:=(-A\cap B)\cup(A\cap -B)$ the *symmetric difference* of A and B. If $A\in \mathcal{B}(K,\mu)$ then cl int(A)$\in \mathcal{B}(K,\mu)$. Moreover $A\Delta(cl\ int(A))$ $\in Z(K,\mu)$. Hence $\mathcal{B}(K,\mu)=\{A\subset K \mid cl\ int(A)$ is a μ-continuity set and $A\Delta(cl\ int(A))$ is a μ-Jordan zero set$\}$.

$\mathcal{B}(K,\mu)$ contains the class $rc_o(K,\mu)$ of *all regular closed, non void μ-continuity sets* which forms a neighbourhood basis system of K (see [15] for an explicit proof).

In general it is sufficient to consider subfields of $\mathcal{B}(K,\mu)$ containing a neighbourhood basis system of K: Let $\mathbb{B}(K,\mu)\subset \mathcal{B}(K,\mu)$ be a field such that $\mathbb{B}(K,\mu)$ contains a neighbourhood basis system of K. Then $\mathbb{B}(K,\mu)$ is called a *μ-basis field.* If $\mathbb{B}(K,\mu)$ fulfills additionally the condition '$A\in\mathbb{B}(K,\mu)$ implies $cl(A)\in\mathbb{B}(K,\mu)$', then $\mathbb{B}(K,\mu)$ is called an *integral μ-basis field.* Let $A\in \mathcal{B}(K,\mu)$, let $\varepsilon>0$, and let $\mathbb{B}(K,\mu)$ be any μ-basis field. Then there is a $\mathbb{B}\in B(K,\mu)$ with $\mu(cl(A\Delta B))<\varepsilon$.

<u>EXAMPLE A.2.1.</u> (A) Let K be the unit interval [0,1] and let μ be the Lebesgue measure λ restricted onto [0,1]. Then $\mathcal{B}([0,1],\lambda)$ is the class of Jordan measurable sets. With respect to the classical calculus it suffices to consider the field $RA([0,1],\lambda)$ generated by all intervalls having rational endpoints. (B) Let K be a Stone space and let μ

be any non negative Radon measure on K. Then the field of all clopen subsets of K is an integral μ-basis field.

Let $\overline{\mathcal{B}}(K):=\{A\Delta N \mid A\in \mathcal{B}(K)$ and N is a subset of a Borel-μ-zero set$\}$ and let $\bar{\mu}: \overline{\mathcal{B}}(K)\to[0,1]$ be defined by $\bar{\mu}(A\Delta N)=\mu(A)$ where A and N are as above. Then $\overline{\mathcal{B}}(K)$ is a σ-field and $\bar{\mu}$ is a normalized, non negative σ-additive measure. $\bar{\mu}$ is *complete* i.e. if $A\in \overline{\mathcal{B}}(K)$, $B\subset A$, and $\bar{\mu}(A)=0$ then $B\in \overline{\mathcal{B}}(K)$. The pair $(\overline{\mathcal{B}}(K),\bar{\mu})$ is the *completion* of μ defined on $\mathcal{B}(K)$. Observe that $\mathcal{B}(K,\mu)\subset \overline{\mathcal{B}}(K)$ whereas $\mathcal{B}(K,\mu)$ is in general not contained in $\mathcal{B}(K)$.

The restriction μ^c of $\bar{\mu}$ onto $\mathcal{B}(K,\mu)$ is called the μ-*content* or the μ-*Jordan content* on K. Observe that $\mu^c(A)=\mu(cl(A))=\mu(cl\ int(A))$ for $A\in \mathcal{B}(K,\mu)$. The μ-content μ^c is complete, non negative, normalized, and finitely additive. In general μ^c is not σ-additive. The restriction $\mu^{\mathbb{B}}$ of $\bar{\mu}$ onto a μ-basis field $\mathbb{B}(K,\mu)$ is called by us the μ-$\mathbb{B}$-*content*.

<u>EXAMPLE A.2.2</u> We consider the Lebesgue measure λ on the unit interval $[0,1]$. Then λ^c is the ordinary Jordan content and λ^{RA} is its restriction onto the field $RA([0,1],\lambda)$ generated by all rational intervals.

A characterizing property of the Jordan content and the Riemann-Darboux integration theory is the possibility to treat them equivalently on dense subsets of the considered compact space:

Let S be a dense subset of K. We define $\mathcal{B}(K,\mu,S):=\mathcal{B}(K,\mu)\cap S$ and $\mu_S^c: \mathcal{B}(K,\mu,S)\to[0,1]$ by $\mu_S^c(A)=\mu(cl(A))$. Then μ_S^c is a non negative, normalized, finitely additive measure defined on the field $\mathcal{B}(K,\mu,S)$. We call μ_S^c the (induced, topological) μ-*trace content*. If $\mathbb{B}(K,\mu)$ is a μ-basis field then we obtain analogously the μ-$\mathbb{B}$-*trace content* $\mu_S^{\mathbb{B}}$ on $\mathbb{B}(K,\mu,S)$.

<u>EXAMPLE A.2.3</u> We consider a set F and a field $\mathfrak{F}$ on F i.e. $\mathfrak{F}$ is a Boolean subalgebra of $\mathbb{P}(F)$. Assume that $\mathfrak{F}$ *separates* F i.e. that for every two elements $a,b\in F$, $a\neq b$ there is an $A\in\mathfrak{F}$ with $a\in A$ and $b\notin A$. Then F interpreted as a set of fixed ultrafilters is a *dense* subset of Stone$(\mathfrak{F})$.

(If $\mathfrak{F}$ separates F not then consider the equivalence relation ~ on F defined by a ~b iff $A\in\mathfrak{F} \Rightarrow a,b\in A$ or $a,b\notin A$. Then $\mathfrak{F}/\sim$ separates F/~ and $\mathfrak{F}/\sim$ is Boolean isomorphic to $\mathfrak{F}$.) Let $r^{CO}:\mathfrak{F}\to[0,1]$ be a (non negative, normalized) finitely additive measure. In analogy to part A.1.3 r^{CO} extends uniquely to a the non negative Radon measure r defined on Stone($\mathfrak{F}$). Then $\mathfrak{F}$ can be seen as the CO-basis field on Stone($\mathfrak{F}$) consisting of all clopen subsets of Stone($\mathfrak{F}$). Hence we obtain the original r^{CO} as the induced topological r-CO-trace content r_F^{CO}. Consequently, *every (bounded, non negative) finitely additive measure defined on a field can be interpreted as a trace content.*

It remains to discuss the equivalence indicated above. The different concepts of contents considered within this introductory treatment are characterized in a special way by the unique, finitely additive measure algebra $\mathfrak{A}(K,\mu):=\mathfrak{B}(K,\mu)/Z(K,\mu)$. A discussion of this unifying approach implies the use of Boolean quotient algebras which we could not introduce within the elementary first part of the appendix. Moreover such a necessarily deeper investigation would offend the supporting nature of this appendix. Therefore the reader should allow us to refer him to our grounding treatment [16]. The equivalence mentioned above will be indicated later on by different methods of calculating integrals yielding identical results.

A.2.2 <u>THE GENERALIZED RIEMANN INTEGRAL</u>. The definition of the generalized Riemann integral coincides widely with its classical counterpart known from high-school or undergraduate courses. The only problem waiting for a solution remains the meaning of an interval in an arbitrary compact space. We will realize soon that the answer was given in the previous section. Similarly as before less known properties of the old integration tool are presented.

Let K be a compact (Hausdorff-)space endowed with a non negative, normalized Radon measure μ. Since an integral calculus is meaningfull only on the *support* $\mathrm{supp}(\mu):=\{x\in K \mid \mu(U(x))>0$ holds for every open neighbourhood $U(x)$ of $x\}$ of μ we *assume* in the following that $K=\mathrm{supp}(\mu)$.

In the classical case the definition of the Riemann integral is performed by partitioning the unit interval in closed subintervals. Here partitions of K are used consisting of elements of $rc_o(K,\mu)$, i.e. consisting of non void, regular closed μ-continuity subsets of K. Hence, *non void, regular closed* μ-*continuity subsets* of K *replace* the notion of *closed subintervals* used by B.Riemann within his integration tool:

Let $\mathbb{B}(K,\mu)$ be an integral μ-basis field. Since $cl(A)\in\mathbb{B}(K,\mu)$ for $A\in\mathbb{B}(K,\mu)$ it follows that $\mathbb{B}(K,\mu)$ contains a neighbourhood basis system of K consisting of elements of $rc_o(K,\mu)$. A $\mathbb{B}(K,\mu)$-*partition* of K is a finite class $\mathcal{P}:=\{P_i\}_{i=1}^n\subset\mathbb{B}(K,\mu)\cap rc_o(K,\mu)$ with $\cup_{i=1}^n P_i=K$ and $\mu(P_i\cap P_j)=0$ for $i\neq j$. Let $\mathcal{P}':=\{P_j'\}_{j=1}^m$ a further $\mathbb{B}(K,\mu)$-partition. Then $\mathcal{P}'$ is *finer* than $\mathcal{P}$ - denoted $\mathcal{P}'\geq\mathcal{P}$ - if for every P_i there is a subset $p(i)\subset\{1,\dots,m\}$ with $P_i=\cup_{j\in p(i)}P_j'$. The $\mathbb{B}(K,\mu)$-partition $\mathcal{P}\cap\mathcal{P}':=\{cl\ int(P_i\cap P_j') \mid 1\leq i\leq n,\ 1\leq j\leq m\}\setminus\{\emptyset\}$ is always finer than both $\mathcal{P}$ and $\mathcal{P}'$. Therefore the class $\mathbb{SB}\mathcal{P}$ of all $\mathbb{B}(K,\mu)$-partitions is *directed* with respect to the order $\geq$.

Let $f:K\rightarrow\mathbb{R}^n$ be a *bounded* function. Let $\mathcal{P}=\{P_i\}_{i=1}^n\in\mathbb{SB}\mathcal{P}$ be a $\mathbb{B}(K,\mu)$-partition of K. If $x_i\in P_i$ then $x:=\sum_{i=1}^n f(x_i)\mu(P_i)$ is called a $\mathcal{P}$-*sum* of f. We denote $S(f,\mathcal{P}):=\{x\in\mathbb{R}^n \mid x$ is a $\mathcal{P}$-sum of $f\}$. If the indexed class $\{S(f,\mathcal{P}) \mid \mathcal{P}\in(\mathbb{SB}\mathcal{P},\geq)\}$ is converging in $\mathbb{R}^n$ with respect to the directed class $(\mathbb{SB}\mathcal{P},\geq)$ then f is called μ-*Riemann integrable* and

$$^R\!\!\int_K f d\mu := \lim_{\mathcal{P}\in(\mathbb{SB}\mathcal{P},\geq)} S(f,\mathcal{P})$$

is called the μ-*Riemann integral* of f over K. Within this context remember the meaning of the convergence of $\{S(f,\mathcal{P}) \mid \mathcal{P}\in(\mathbb{SB}\mathcal{P},\geq)\}$: If $\varepsilon>0$ then there is a $\mathcal{P}\in\mathbb{SB}\mathcal{P}$ such that the following holds: If $\mathcal{P}'\geq\mathcal{P}$, $\mathcal{P}''\geq\mathcal{P}$, and $x'\in S(f,\mathcal{P}')$, $x''\in S(f,\mathcal{P}'')$, then $\|x'-x''\|<\varepsilon$ where $\|.-.\|$ denotes the Euclidean distance.

Since the μ-Riemann integral is a limit, all operations compatible with limits are compatible with the above integral. In particular, if $A \in rc_o(K,\mu)$ then we can integrate over A only. Moreover the above definition is independent of the chosen integral μ-basis field.

As well as in the classical case the μ-Riemann integrability can be characterized by Darboux' lower - and Darboux' upper sums. For the sake of simplicity (and generality) we combine here both directly:

If $M \subset \mathbb{R}^n$ then $dia(M) := \sup\{ \|x-y\| \mid x,y \in M\}$ denotes the *diameter* of M with respect to the Euclidean distance. Then $dis(f,\mathcal{P}) := \sum_{i=1}^{n} dia(f(P_i))\,\mu(P_i)$ denotes the μ-*distance sum* of f with respect to the $\mathbf{B}(K,\mu)$-partition $\mathcal{P}$. Then f is μ-*Darboux integrable* iff $\inf\{dis(f,\mathcal{P}) \mid \mathcal{P} \in \mathbb{SB}\,\mathcal{P}\}=0$. Following B.Riemann's original proof idea one deduces that f *is* μ-*Riemann integrable iff* f *is* μ-*Darboux integrable*. It is worth mentioning that the situation is different for Banach space valued functions. Moreover, observe that the above definition works as well for a function $g: K \longrightarrow P$ where P is the compact, metric space of preferences.

<u>PROPOSITION A.1.</u> Let K be a compact space and let μ be a non negative Radon measure with $supp(\mu)=K$. Let $f: K \to \mathbb{R}^n$ be a bounded mapping.
(a) f is μ-Darboux integrable iff f is continuous μ-almost everywhere (i.e. the set of discontinuity points of f is contained in a μ-zero set).
(b) If f is μ-Darboux integrable then f is $\bar{\mu}$-Lebesgue integrable and its μ-Riemann integral coincides with its $\bar{\mu}$-Lebesgue integral. Here $\bar{\mu}$ denotes the completion of μ.

The proof of (b) is strictly following the classical proof idea. We restrict ourselves to the proof of (a) which requires some new notions:

A point $x \in K$ is called an a-*discontinuity point* of f provided that in every neighbourhood U(x) of x there are $y,z \in U(x)$ with $\|f(z)-f(y)\| \geq a$. We denote DC(f,a) the closed set of all a-discontinuity points of f.

Then (a) can be reformulated equivalently as follows: (a') f *is* μ*-Darboux integrable iff* $\mu(DC(f,a))=0$ *for every* $a>0$. Therefore it suffices to prove (a'):

If f is μ-Darboux integrable then we obtain $\mu(DC(f,a))=0$ from the definition of the μ-Darboux integrability: Since f is μ-Darboux integrable there is a sequence $\{\mathcal{P}_i\}_{i=1}^{\infty}$ of $\mathbb{B}(K,\mu)$-partitions of K with $dis(f,\mathcal{P}_i)\to 0$ for $i\to\infty$. If $\mathcal{P}_i=\{P_{ij}\mid 1\le j\le n(i)\}$ then $Z:=\bigcup_{i,j}\partial P_{ij}$ fulfills $\mu(Z)=0$. Hence $\mu(DC(f,a)\cap Z)=0$. On the other hand $\mu(DC(f,a)\cap(K\setminus Z))=0$ because $\inf_i dis(f,\mathcal{P}_i)=0$.

Assume that $\mu(DC(f,a))=0$ for every $a>0$. Let $\varepsilon>0$ and let $F\in\mathbb{R}_+$ with $\|f(x)\|<F$ for every $x\in K$. Then there is a regular closed μ-continuity set A with $DC(f,\varepsilon/2)\subset int(A)$ and $\mu(A)<\varepsilon/(2F)$. For every $x\in cl(K\setminus A)$ we can choose a regular closed μ-continuity set $U(x)$ with $x\in int(U(x))$ and $dia(f(U(x)))<\varepsilon/2$. We can assume that $U(x)\cap DC(f,\varepsilon/2)=\emptyset$. There are finitely many x_i, $i=1,\dots,n$ such that $\{int(U(x_i))\}_{i=1}^{n}$ is an open covering of $cl(K\setminus A)$. Let $B:=K\setminus\bigcup\{int(U(x_i))\mid 1\le i\le n\}$. Then B together with $\{U(x_i)\}_{i=1}^{n}$ yields a partition $\mathcal{P}$ of K consisting of non void, regular closed μ-continuity subsets of K, such that $dis(f,\mathcal{P})\le F(\varepsilon/2F)+(\varepsilon/2)\mu(K)\le\varepsilon$. Here we supposed without loss of generality $\mu(K)=1$. Hence f is μ-Darboux integrable. Q.E.D.

<u>EXAMPLE A.2.4.</u> (a) Let K be the unit interval [0,1] and let μ be the Lebesgue measure λ restricted on [0,1]. Then we obtain the classical Riemann integration tool. (b) Let $K=\gamma\mathbb{N}\cong\mathbb{N}\cup\{\infty\}$ from example A.1.3-A.1.4. If $i\in\mathbb{N}\cup\{\infty\}$ then δ_i denotes the normalized Dirac measure centered at i; i.e. $\delta_i(A)=1$ iff $i\in A$. Let $\mu:=(1/2)\delta_\infty+\sum_{i=1}^{\infty}(1/2^{i+1})\delta_i$. Then μ is a normalized, non negative Radon measure on $\gamma\mathbb{N}$ with $supp(\mu)=\gamma\mathbb{N}$. A bounded function $f:\gamma\mathbb{N}\to\mathbb{R}^n$ is μ-Darboux integrable iff f is continuous in ∞, i.e. iff $\lim_{n\to\infty} f(n)=f(\infty)$.

The integral calculus introduced above can be treated equivalently on every dense subset S of K by considering $\mathbb{B}(K,\mu,S)$ and $\mu_S^{\mathbb{B}}$. Naturally prop. A.1. (a), (b) can not be applied. But the condition (a') is not affected by the alteration and characterizes P-valued functions as well. Moreover the result of the calculated integral remains unchanged. This kind of stability of the Riemann Darboux integration is important with respect to our grounding axiom (A5). With respect to applications the following, more specialized approach based on J.v.NEUMANN's rearrangement theorem [A7] is probably more instructive:

We consider a compact (Hausdorff-)space K endowed with a normalized, non negative Radon measure μ such that supp(μ)=K. Let $\{x_i\}_{i=1}^{\infty}$ be a sequence in K. Then $\{x_i\}_{i=1}^{\infty}$ is called μ-*uniformly distributed* iff

$$\lim_{n\to\infty} \frac{1}{n} \sum_{i=1}^{n} f(i) = \int_K f d\mu$$

holds for every continuous, real valued function f on K.

If we assume additionally K to be *metrizable* then μ-uniformly distributed sequences are not scarce: (i) We interpret a sequence $x := \{x_i\}_{i\in\mathbb{N}} \subset K$ as an element of the countable product $K_\infty := \Pi_1^\infty K$. With respect to the canonical product measure $\mu_\infty := \Pi_1^\infty \mu$ almost all sequences x are uniformly distributed (see L.KUIPERS and H.NIEDERREITER [24] chapter 3 theorem 2.2). (ii) Let us restrict to the case that μ is *atomless*. Then supp(μ) contains no isolated point and we obtain: Let S be a countable dense subset of K. Then S can be enumerated in such a way that we obtain a μ-uniformly distributed sequence (see L.KUIPERS and H.NIEDERREITER [24] chapter 3 theorem 2.5).

PROPOSITION A.2 Let K be a compact space and let μ be a normalized, non negative Radon measure with supp(μ)=K. Let $\{x_i\}_{i=1}^{\infty}$ be a μ-uniformly distributed sequence in K. Then

$$^R\int_K f d\mu = \lim_{n\to\infty} \frac{1}{n} \sum_{i=1}^{n} f(i)$$

holds for every bounded, μ-Darboux integrable, $\mathbb{R}^n$-valued function on K.

For an explicit proof we refer the reader to [15] of S.ROLEWICZ and the author. With respect to applications of the above proposition in numerical analysis we refer the reader to E.HLAWKA [14], part VIII, as well as to L.KUIPERS and H.NIEDERREITER [24] chapter 2.5.

A.2.3. THE SPACE $D(K,\mu,\mathbb{R}^n)$. With respect to the classical approach the material presented here is not new (cf. K.G.JOHNSON [A4], Z.SEMADENI [33]) but nevertheless widely unfamiliar because it is seldom treated in analysis courses. Therefore it seems to us to be reasonable to give an explicit introduction of the space $D(K,\mu,\mathbb{R}^n)$ of μ-Darboux integrable functions which is a closed subspace of the space $L_\infty(K, \overline{\mathcal{B}}(K), \overline{\mu},\mathbb{R})$ of μ-essentially bounded, $\overline{\mathcal{B}}(K)$-measurable functions.

We denote $\mathcal{D}(K,\mu,\mathbb{R}^n):=\{f:K \to \mathbb{R}^n \mid f$ is a bounded, μ-Darboux integrable function$\}$. Then $\mathcal{D}(K,\mu,\mathbb{R}^n)$ is a linear space. If $n=1$, then it is an algebra. We introduce on $\mathcal{D}(K,\mu,\mathbb{R}^n)$ the *Darboux semi norm* $\|\ \|_D:\mathcal{D}(K,\mu,\mathbb{R}^n) \to \mathbb{R}_+$ as follows:

$$(*) \qquad \|f\|_D:=\sup_{A\in rc_0(K,\mu)} \inf_{x\in A} \|f(x)\|.$$

Then $\|\ \|_D$ *is a semi norm* on $\mathcal{D}(K,\mu,\mathbb{R}^n)$ (i.e. the implication $\|f\|_D=0 \Rightarrow f\equiv 0$ is not necessarily fulfilled. The other axioms of a norm remain true: (i) $f\equiv 0 \Rightarrow \|f\|_D=0$. (ii) $\|\alpha f\|_D=|\alpha|\ \|f\|_D\ \forall\alpha\in\mathbb{R}$. (iii) $\|f+g\|_D \leq \|f\|_D+\|g\|_D$). Instead of the formula (*) one can use equivalently the following ones:

$$(**) \qquad \|f\|_D=\sup_{A\in rc_0(K,\mu)} \frac{1}{\mu(A)}\ {}^R\!\!\int_A \|f\|\,d\mu$$

$$(***) \qquad \|f\|_D=\sup\{\|f(x)\| \mid f \text{ is continuous in } x\}.$$

Remind the definition of the μ-ess. sup (semi) norm: $\|g\|_{ess}:=\inf\{c\in\mathbb{R} \mid \|g(x)\| \leq c$ for all $x\in K$ except on a μ-zero set$\}$ where g is a μ-essentially bounded, $\overline{\mathcal{B}}(K)$-measurable, $\mathbb{R}^n$-valued function on K. We recognize by

(***) and prop. A.1 that $\| \|_D$ *equals* $\| \|_{ess}$ *on* $\mathcal{D}(K,\mu,\mathbb{R}^n)$. Naturally, if arbitrary bounded, $\overline{\mathcal{B}}(K)$-measurable functions are considered, then $\| \|_D$ is different from $\| \|_{ess}$ in general.

The Darboux semi norm can be equivalently calculated with respect to all non void, regular closed μ-continuity sets contained in an integral μ-basis field $\mathbf{B}(K,\mu)$. Moreover $\| \|_D$ can be applied analogously with respect to a dense subset S of K. If $\mathbb{R}^n$ is replaced by a compact metric space then we obtain in this way a semi metric instead of a semi norm. Lastly, if n=1 then the following condition $\|fh\|_D \leq \|f\|_D \|h\|_D$ is fulfilled for $f,h \in \mathcal{D}(K,\mu,\mathbb{R})$.

EXAMPLE A.2.5 (a) Let $\{q_i\}_{i\in\mathbb{N}}$ be a sequence consisting of all rational numbers contained in the unit interval (0,1). We define $f:[0,1] \to \mathbb{R}$ by $f(q_i):=i^{-1}$ and $f(x):=0$ otherwise. Then f is Riemann-, resp. λ-Darboux-, integrable and $\|f\|_D = \|f\|_{ess} = 0$. (b) Let $g:[0,1] \to \mathbb{R}$ be defined by $g(x):=1$ iff x is irrational and $g(x):=0$ otherwise. Then g is Lebesgue integrable but not Riemann-, resp. λ-Darboux-, integrable. Moreover $\|g\|_D = 0$ whereas $\|g\|_{ess} = 1$.

We denote $Z\mathcal{D}(K,\mu,\mathbb{R}^n) := \{f \in \mathcal{D}(K,\mu,\mathbb{R}^n) \mid \|f\|_D = 0\}$. Then $Z\mathcal{D}(K,\mu,\mathbb{R}^n)$ is a closed subspace of $\mathcal{D}(K,\mu,\mathbb{R}^n)$. Observe that $\|f\|_D = 0$ iff ${}^R\!\int_K \|f\| d\mu = 0$ - provided that $f \in \mathcal{D}(K,\mu,\mathbb{R}^n)$. We denote by $D(K,\mu,\mathbb{R}^n)$ the following *quotient space*: $D(K,\mu,\mathbb{R}^n) := \mathcal{D}(K,\mu,\mathbb{R}^n)/Z\mathcal{D}(K,\mu,\mathbb{R}^n)$. In the case n=1 we obtain the closed ideal $Z\mathcal{D}(K,\mu,\mathbb{R})$ and $D(K,\mu,\mathbb{R})$ turns out to be an algebra. Moreover the above construction can be applied analogously for P-valued functions. Since $\| \|_D$ and $\mathcal{D}(K,\mu,\mathbb{R}^n)$ - resp. $\mathcal{D}(K,\mu,P)$ - can be considered equivalently on a dense subset S of K this holds true for $D(K,\mu,\mathbb{R}^n)$ - resp. $D(K,\mu,P)$ - as well.

It is well known that the μ-ess. sup. semi norm induces a norm on the space $L_\infty(K,\overline{\mathcal{B}}(K),\bar{\mu},\mathbb{R}^n)$ consisting of $\overline{\mathcal{B}}(K)$-measurable, $\bar{\mu}$-essential bounded functions modulo functions with a zero integral (see e.g. N.DUNFORD and J.T.SCHWARTZ [A1]). In a strict analogy the Darboux semi norm induces on $D(K,\mu,\mathbb{R}^n)$ a norm. We call this norm the *Darboux norm* denoted $\| \|_D$ too. With respect to this norm $D(K,\mu,\mathbb{R}^n)$ is a normed linear space, $D(K,\mu,\mathbb{R})$ is a normed algebra, and $D(K,\mu,P)$ turns out to be a metric space. With a somewhat technical proof (see [16]) one can show even more:

PROPOSITION A.3 Let K be a compact space. Let μ be a non negative Radon measure on K with $supp(\mu)=K$. Then $(D(K,\mu,\mathbb{R}^n), \| \|_D)$ is a Banach space, $(D(K,\mu,\mathbb{R}), \| \|_D)$ is a Banach algebra, and $(D(K,\mu,P), \| \|_D)$ is a complete metric space.

The complete spaces introduced above require the handling of equivalence classes of functions. For practical purposes it may be more instructive to work instead with typical representatives. We want to show here a way how such functions can be singled out of equivalence classes:

Remember that $DC(f,a):=\{x\in K|$ in every neighbourhood $U(x)$ of x there are y,z with $\|f(y)-f(z)\| \geq a\}$ is the set of all a-*discontinuity points* of f. We denote $C(f)$ the set of all *continuity points* of f and $DC(f):= \cup_{a>o} DC(f,a)$ denotes the set of all discontinuity points of f. If $x\in DC(f)$ and if for every $\varepsilon>0$ there exists a neighbourhood $U_\varepsilon(x)$ of x such that $\|f(z)-f(y)\|<\varepsilon$ for $z,y\in U_\varepsilon(x)\cap C(f)$ then x is called an *inessential discontinuity point* of f. We denote iness $DC(f)$ the set of all inessential discontinuity points of f. If for every neighbourhood $U(x)$ of x there are $y,z\in U(x)\cap C(f)$ with $\|f(y)-f(z)\| \geq a$ then x is called an a-*essential discontinuity point* of f. We denote ess $DC(f,a)$ the set of all a-essential discontinuity points of f and ess $DC(f):=\cup_{a>o}$ ess $DC(f,a)$.

We obtain the disjoint decompositions $K=C(f)\cup DC(f)=C(f)\cup$ess $DC(f)$ $\cup$iness $DC(f)$. The restriction $f|_{C(f)}$ can be extended continuously onto $C(f)\cup$iness $DC(f)$. If $a>0$ then ess $DC(f,a)\cup$iness $DC(f,a) \subset DC(f,a)$. If iness $DC(f,a)=\emptyset$ and if ess $DC(f,a)=DC(f,a)$ holds for every $a>0$ then f is called *polished*. We denoted $\|f\|_o:=\{\sup \|f(x)\| \; |x\in K\}$ the sup norm of f. If f is polished then $\|f\|_o=\|f\|_D=\|f\|_{ess}$.

If $h \in D(K,\mu,\mathbb{R}^n)$ then there exists obviously a polished representative $f \in \mathcal{D}(K,\mu,\mathbb{R}^n)$ of the equivalence class h. We deem polished functions to be easy enough to handle. From our viewpoint they allow us to consider elements of $D(K,\mu,\mathbb{R}^n)$ as 'broken-continuous' functions.

<u>EXAMPLE A.2.6</u> (a) Again, let $\{q_i\}_{i\in\mathbb{N}}$ be a sequence consisting of all rational numbers contained in the unit interval (0,1). We define $f:[0,1] \to \mathbb{R}$ by $f(q_i)=i^{-1}$ and $f(x)=0$ otherwise. The unique polished representative of the equivalence class determined by f in $D([0,1],\lambda,\mathbb{R})$ is the function identical null. (b) Let $f:[0,1] \to \mathbb{R}$ be defined by $f(x)=0$ on $[0,1/2)$, $f(1/2)=2$, and $f(x)=1$ on $(1/2,1]$. Then 1/2 is a 2-discontinuity point of f and 1-essential discontinuity point of f. Then $g: [0,1] \to \mathbb{R}$ defined by $g(1/2)=1$ and $g\equiv f$ otherwise is a polished representative of the equivalence class determined by f in $D([0,1],\lambda,\mathbb{R})$.

It remains to present a deeper and principal characterization of Banach spaces of Darboux integrable functions. Naturally we are unable to give a proof within this introductory treatment. For a verification we refer the reader to [16] and [17].

Remind that $\mathfrak{A}(K,\mu)$ denoted the Boolean quotient algebra $\mathfrak{B}(K,\mu)/Z(K,\mu)$. In analogy to part A.1.3 μ induces a Radon measure on $\mathrm{Stone}(\mathfrak{A}(K,\mu))$ such that the following holds:

<u>PROPOSITION A.4</u> Let K be a compact space. Let μ be a non negative Radon measure with $\mathrm{supp}(\mu)=K$. Then $(D(K,\mu,\mathbb{R}^n), \|\ \|_D)$ is isometrically isomorphic to $(C(\mathrm{Stone}(\mathfrak{A}(K,\mu)),\mathbb{R}^n), \|\ \|_0)$ such that the Riemann integral is not offended. Here C(,) denotes the space of continuous functions.

By the above proposition one can recognize that measure preserving Boolean isomorphisms characterize isometrically isomorphic spaces of Darboux integrable functions.

A3 SOME PROOFS

Within this third part of the appendix we want to present some technical proofs to improve the selfcontainedness of our research paper.

A.3.1 PROOFS CONCERNING CHAPTER I.

In this section we reprove three results refered only in the first chapter. At first the necessary, technical notations shall be remembered. (For a more general approach we refer the reader to [20] and [21]):

Let $r>0$ and $m\in\mathbb{N}:=\{1,2,3...\}$. If $i\in\{0,1,2,...,m-1\}$ then S_i is the interval $[ir/m,(i+1)r/m)$ closed to the left and open to the right. Let $s_0=0$ and $s_i=ir/m+r/(2m)$ if $0<i<m$. We call $\alpha=\alpha(r,m):=(\{S_i\}_{0\le i<m},\{s_i\}_{0\le i<m},r)$ a *scale observation of* $\mathbb{R}_+$. We denote $gr(\alpha):=\{s_i\}_{0\le i<m}$ the *grid*, $r(\alpha):=r$ the *range* and $grc(\alpha)=r/m$ the *grid constant* with respect to α.

By a product construction we get canonically the generated scale observation α^n of $\mathbb{R}^n_+$: Let $\xi:=(i_1,...,i_n)$ be a *multi-index* with $0\le i_j<m$ for $1\le j\le n$. Then we denote $S_\xi:=S_{i_1}\times S_{i_2}\times...\times S_{i_n}$ and $s_\xi:=(s_{i_1},s_{i_2},...,s_{i_n})$. Let $\Xi:=\{(i_1,...,i_n)\,|\,0\le i_j<m\ 1\le j\le n\}$ the set of all these multi-indices. Then $\alpha^n=\alpha^n(r,m):=(\{S_\xi\}_{\xi\in\Xi},\{s_\xi\}_{\xi\in\Xi},r)$ is the *scale observation of* $\mathbb{R}^n_+$ generated by α. The related *grid* $\{s_\xi\}_{\xi\in\Xi}$ consists of m^n grid points. The *range* $r(\alpha^n):=r=r(\alpha)$ and the *grid constant* $grc(\alpha^n):=r/m=grc(\alpha)$ remain unchanged by definition.

We denote $\mathbb{R}^n_+:=\{x\in\mathbb{R}^n\,|\,x_i\ge 0 \text{ for } 1\le i\le n\}$. A *preference* on $\mathbb{R}^n_+$ is a subset $\succ\subset\mathbb{R}^n_+\times\mathbb{R}^n_+$ fulfilling the following conditions: (a) $\succ$ is *continuous* i.e. $\succ$ is an open subset of $\mathbb{R}^n_+\times\mathbb{R}^n_+$. A subset $M\subset\mathbb{R}^n_+\times\mathbb{R}^n_+$ is called *open* iff there is an open set $M'\subset\mathbb{R}^n\times\mathbb{R}^n$ with $M'\cap\mathbb{R}^n_+\times\mathbb{R}^n_+=M$. (b) $\succ$ is *irreflexive* i.e. $(x,x)\notin\succ$ for every $x\in\mathbb{R}^n_+$. (c) $\succ$ is *transitive* i.e. $(x,y)\in\succ$, $(y,z)\in\succ$ imply $(x,z)\in\succ$ for $x,y,z\in\mathbb{R}^n_+$. The *set of all preferences* on $\mathbb{R}^n_+$ is denoted by P.

Let $\succ\subset\mathbb{R}^n_+\times\mathbb{R}^n_+$ be a preference. Let $\alpha:=(\{S_i\}_{0\le i<m},\{s_i\}_{0\le i<m},r)$ be

a scale observation of $\mathbb{R}_+$ and let $\alpha^n:=(\{S_\xi\}_{\xi\in\Xi},\{s_\xi\}_{\xi\in\Xi},r)$ be the generated scale observation of $\mathbb{R}_+^n$. We denote $A(\succ,\alpha^n):=\cup\{S_\xi\times S_{\xi'}|(s_\xi,s_{\xi'})\in\succ\}$ $\subset \mathbb{R}_+^n\times\mathbb{R}_+^n$. We define $\succ(\alpha^n):=\text{int cl}(A(\succ,\alpha^n))$ with respect to $\mathbb{R}_+^n\times\mathbb{R}_+^n$; i.e. $\succ(\alpha^n)$ is the interior of the closure of $A(\succ,\alpha^n)$ where the interior and the closure is formed with respect to the topological space $\mathbb{R}_+^n\times\mathbb{R}_+^n$. Then $\succ(\alpha^n)$ is the *observed preference* with respect to $\succ$ and α^n.

PROPOSITION A.5 Let $\succ\subset\mathbb{R}_+^n\times\mathbb{R}_+^n$ be a preference. Let $\alpha^n=\alpha^n(r,m)$ be a scale observation of $\mathbb{R}_+^n$ generated by the scale observation $\alpha=\alpha(r,m)$ of $\mathbb{R}_+$. The observed preference $\succ(\alpha^n)$ with respect to $\succ$ and α^n is a preference; i.e. $\succ(\alpha^n)\in P$.

P r o o f: $\succ(\alpha^n)\subset\mathbb{R}_+^n\times\mathbb{R}_+^n$. By definition $\succ(\alpha^n)$ is open, hence continuous. By the special construction of the scale observations used here it follows that $\succ(\alpha^n)=\text{int}(A(\succ,\alpha^n))$. Therefore $\succ(\alpha^n)$ is irreflexive and transitive since $\succ$ possesses these two properties. Q.E.D.

Observe that the observed preference introduced above is always *regular open* by definition; i.e. $\succ(\alpha^n)=\text{int}(\text{cl}(\succ(\alpha^n)))$.

A preference $\succ\subset\mathbb{R}_+^n\times\mathbb{R}_+^n$ is called *monotonic* if $\succ$ fulfills the following condition: If $y>x$ - i.e. if $x,y\in\mathbb{R}_+^n$, $x_i\le y_i\,\forall i$, and $x\neq y$ - then y is prefered to x: $(y,x)\in\succ$. The set of all monotonic preferences is denoted P_{mo}.

REMARK A.6 If $\succ\subset\mathbb{R}_+^n\times\mathbb{R}_+^n$ is a monotonic preference then $\succ$ is regular open.

P r o o f: Let $\succ\subset\mathbb{R}_+^n\times\mathbb{R}_+^n$ be a monotonic preference. Since $\succ$ is open we obtain $\succ\subset\text{int cl}(\succ)$. If $(y,x):=((y_1,\ldots,y_n),(x_1,\ldots,x_n))\notin\succ$ and if $(y',x'):=((y_1',\ldots,y_n'),(x_1',\ldots,x_n'))$ with $y_i'<y_i, x_i'>x_i$ for $1\le i\le n$, then $(y',x')\notin\succ$ for $\succ$ is monotonic; i.e. $(y',x')\in\succ$ implies $(y,x)\in\succ$ which yields a contradiction. Consequently, if $(y,x)\in\text{cl}(\succ)\setminus\succ$ then $(y,x)\in$ cl int$(\mathbb{R}_+^{2n}\setminus\succ)$. Hence $(y,x)\notin\text{int cl}(\succ)$. This implies $\succ=\text{int cl}(\succ)$. Q.E.D.

The space P of preferences is endowed with the metric δ yielding the topology of closed convergence. We introduce this well established tool here once more in some detail. At first we collect the related technical framework:

The set $\gamma\mathbb{R}_+^{2n} := \mathbb{R}_+^{2n} \cup \{\infty\}$ consists of the set $\mathbb{R}_+^{2n} := \{x \in \mathbb{R}^{2n} \mid x_i \geq 0$ for $1 \leq i \leq 2n\}$ enlarged by the additional point ∞. We define a topology on $\gamma\mathbb{R}_+^{2n}$ by specifying an open neighbourhood basis system: If $x \in \mathbb{R}_+^{2n} \subset \gamma\mathbb{R}_+^{2n}$ then we choose the ordinary open subsets of $\mathbb{R}_+^{2n}$ containing x as a neighbourhood basis of x. If $x=\infty$ then $\mathfrak{U}(\infty) := \{U_n(\infty)\}_{n\in\mathbb{N}}$ with $U_n(\infty) := \{x \in \mathbb{R}_+^{2n} \mid \|x\| > n\} \cup \{\infty\}$ yields a neighbourhood basis system of ∞. Here $\| \ \|$ denotes the Euclidean norm. Then $\gamma\mathbb{R}_+^{2n}$ endowed with the topology described above is called the *one-point* - or Alexandroff - *compactification* of $\mathbb{R}_+^{2n}$. The one point compactification $\gamma\mathbb{R}_+^{2n}$ is a *compact, metrizable space*. A sequence $\{x_i\}_{i\in\mathbb{N}} \subset \mathbb{R}_+^{2n} \subset \gamma\mathbb{R}_+^{2n}$, bounded in $\mathbb{R}_+^{2n}$ with respect to the Euclidean norm, is converging iff it converges in the Euclidean space $\mathbb{R}_+^{2n}$. If $\{x_i\}_{i\in\mathbb{N}} \subset \mathbb{R}_+^{2n} \subset \gamma\mathbb{R}_+^{2n}$ is a sequence such that $\{\|x_i\|\}_{i\in\mathbb{N}}$ is unbounded and increasing, then $x_i \to \infty$.

We denote $S^{2n} := \{x \in \mathbb{R}^{2n+1} \mid \|x\| = 1\}$ the *unit sphere* in $\mathbb{R}^{2n+1}$. Then $NP := (0,0,\dots,0,1)$ is the North Pole and $SP := (0,0,\dots,0,-1)$ is the South Pole of this unit sphere. We denote $S_+^{2n} := \{x \in S^{2n} \mid x = (x_1,\dots,x_{2n},x_{2n+1})$, $x_i \geq 0$ for $1 \leq i \leq 2n\}$ the sector of S^{2n} over $\mathbb{R}_+^{2n}$, abbreviated: the *plus sector* of S^{2n}. Then S_+^{2n} endowed with the metric induced by the Euclidean distance of $\mathbb{R}^{2n+1}$ is a *compact, metric* space.

The *spherical projection* (or also stereographic projection) $p: S_+^{2n} \to \gamma\mathbb{R}_+^{2n}$ is defined by $p(NP) := \infty$ and by $p(x) = p((x_1,\dots,x_{2n},x_{2n+1})) := (y_1,\dots,y_{2n})$ with $y_i := x_i/(1-x_{2n+1})$ for $1 \leq i \leq 2n$ (cf. W.KLINGENBERG [A6] 1.1.3 (ii)). The spherical projection p is a *homeomorphism* from S_+^{2n} onto $\gamma\mathbb{R}_+^{2n}$. We obtain a picture of p as follows: Let $y \in \mathbb{R}_+^{2n}$ and let $l(y)$ be the straight line in $\mathbb{R}^{2n+1}$ between the North Pole NP and the point y. Let x be the point of intersection of $l(y)$ and the unit sphere S^{2n}. Then x is mapped by p onto y.

Let M be a compact, metrizable space. Then a metric d inducing the original topology on M is *uniquely* determined up to a homeomorphism which is uniformly continuous in both directions; i.e. if d' is a further metric on M inducing also the given topology, then there are positive K_1, K_2 with $K_1 d(x,y) \leq d'(x,y) \leq K_2 d(x,y)$ (cf. J.L.KELLEY [A5] chapter 6 theorem 29). We conclude that all metrics on M inducing the original topology are *uniformly equivalent.*

Let M,M' are compact, metrizable spaces. Let $h: M \to M'$ be a homeomorphism. Then h is *uniformly continuous* in both directions (cf. J.L. KELLEY [A5] ibidem).

The spaces $\gamma\mathbb{R}^n_+$ and S^{2n}_+ introduced above are both compact and metrizable. The spherical projection $p: S^{2n}_+ \to \gamma\mathbb{R}^{2n}_+$ is a homeomorphism. Hence p as well as p^{-1} are uniformly continuous. Therefore we can study $S^{2n}_+ = p^{-1}(\gamma\mathbb{R}^{2n}_+)$ *instead of* $\gamma\mathbb{R}^{2n}_+$ without affecting convergent sequences, Cauchy sequences, or the closure operation. In particular, the distance induced on $\gamma\mathbb{R}^{2n}_+$ by p^{-1} and the Euclidean distance of $\mathbb{R}^{2n+1}$ is uniformly equivalent to the Euclidean distance on $\mathbb{R}^{2n}_+$ provided that a compact subset of $\mathbb{R}^{2n}_+$ is considered only.

Let (M,d) be a compact, metric space. Then $\mathbf{F}_o(M)$ denotes the class of all *non void, closed subsets* of M. If $A \in \mathbf{F}_o(M)$ and $x \in M$ then $d(A,x) := \inf\{d(y,x) \mid y \in A\}$. If $\varepsilon > 0$ then $B_\varepsilon(A) := \{x \in M \mid d(A,x) < \varepsilon\}$ denotes the open ε-ball around A. A *metric* $\delta_d: \mathbf{F}_o(M) \times \mathbf{F}_o(M) \to \mathbb{R}_+$ is defined as follows: $\delta_d(A,B) := \inf\{\varepsilon > 0 \mid A \subset B_\varepsilon(B) \text{ and } B \subset B_\varepsilon(A)\}$. The metric δ_d is called the *Hausdorff distance* induced by d on $\mathbf{F}_o(M)$. One can prove that *$(\mathbf{F}_o(M), \delta_d)$ is a compact, metric space* (cf. R.ENGELKING [A2] 3.12.26).

We are prepared now to undertake the final step for introducing a topology on the space P of all preferences on the commodity space $\mathbb{R}^n_+$:

Let $\succ \subset \mathbb{R}^n_+ \times \mathbb{R}^n_+$ be a preference. Then $\succ$ is an open subset of $\mathbb{R}^{2n}_+$. We denote $\overline{\succ} := \gamma\mathbb{R}^{2n}_+ \setminus \succ$. Since $\infty \in \overline{\succ}$ it follows that $\overline{\succ}$ is a closed and non void subset of $\gamma\mathbb{R}^{2n}_+$; i.e. $\overline{\succ} \in \mathbf{F}_o(\gamma\mathbb{R}^{2n}_+)$. Assume that $\succ'$ is a further preference. Then $p^{-1}(\overline{\succ})$ and $p^{-1}(\overline{\succ'})$ are closed, non void subsets of

S_+^{2n}. Let $\| \ \|$ be the Euclidean distance of $\mathbb{R}^{2n+1}$ and let $\delta_{\| \ \|}$ be the Hausdorff distance induced by $\| \ \|$ on $\mathbf{F}_o(S_+^{2n})$. Then $\delta(\succ,\succ'):=\delta_{\| \ \|}(p^{-1}(\bar{\succ}),p^{-1}(\bar{\succ}'))$ is well defined. We call $\delta(\succ,\succ')$ the *closed convergence distance* of $\succ$ and $\succ'$. Since $\delta_{\| \ \|}$ is a metric on $\mathbf{F}_o(S_+^{2n})$ it follows that the closed convergence distance δ is a *metric* on P. The uniformity induced by δ on P is called the *topology of closed convergence.* One can prove that (P,δ) *is a compact, metric space* (cf. W.HILDENBRAND [13] part I B II and part II chapter I 1.2 for an explicit proof).

We want to terminate the depiction of the space of preferences by a helpful technic for calculating limits in (P,δ). Assume that (M,d) is a compact, metric space. Let $\{x_i\}_{i\in\mathbb{N}}$ be a sequence in M. We denote $Li(\{x_i\}_{i\in\mathbb{N}}):=\{y\in M|$ If $U(y)$ is an open neighbourhood of y then $x_i\in U(y)$ holds for all $i\in\mathbb{N}$ except finitely many i$\}$ the *topological limes inferior* and $Ls(\{x_i\}_{i\in\mathbb{N}}):=\{y\in M|$ If $U(y)$ is an open neighbourhood of y then $U(y)\cap \{x_i\}_{i\in\mathbb{N}}$ is not finite$\}$ the *topological limes superior* of $\{x_i\}_{i\in\mathbb{N}}$. Then $Li(\{x_i\}_{i\in\mathbb{N}})$ is the *limit* of $\{x_i\}_{i\in\mathbb{N}}$ provided that $\{x_i\}_{i\in\mathbb{N}}$ is converging or a Cauchy sequence. Otherwise this topological limes inferior is empty. The *accumulation points* of $(\{x_i\}_{i\in\mathbb{N}})$ correspond to $Ls(\{x_i\}_{i\in\mathbb{N}})$. Iff $\{x_i\}_{i\in\mathbb{N}}$ is converging or a Cauchy sequence then $Li(\{x_i\}_{i\in\mathbb{N}})= Ls(\{x_i\}_{i\in\mathbb{N}})$. We apply these notions for sequences in (P,δ):

Let $\{\succ_i\}_{i\in\mathbb{N}}$ be a sequence in P and let $\{\bar{\succ}_i\}_{i\in\mathbb{N}}$ be the related sequence in $\mathbf{F}_o(\gamma\mathbb{R}_+^{2n})$ with $\bar{\succ}_i:=\gamma\mathbb{R}_+^{2n}\setminus\succ_i$. Then $Li(\{\bar{\succ}_i\}_{i\in\mathbb{N}}):=\{y\in\gamma\mathbb{R}_+^{2n}|$ There is a sequence $\{x_i\}_{i\in\mathbb{N}}$ with $x_i\in\bar{\succ}_i \forall i\in\mathbb{N}$ such that $y=Li(\{x_i\}_{i\in\mathbb{N}})\}$ is the *topological limes inferior* and $Ls(\{\bar{\succ}_i\}_{i\in\mathbb{N}}):=\{y\in\gamma\mathbb{R}_+^{2n}|$There is a sequence $\{x_i\}_{i\in\mathbb{N}}$ with $x_i\in\bar{\succ}_i\forall i\in\mathbb{N}$ such that $y=Ls(\{x_i\}_{i\in\mathbb{N}})\}$ is the *topological limes superior* of $\{\bar{\succ}_i\}_{i\in\mathbb{N}}$. Iff $\{\succ_i\}_{i\in\mathbb{N}}$ is convergent or a Cauchy sequence in (P,δ) then $Li(\{\bar{\succ}_i\}_{i\in\mathbb{N}})=Ls(\{\bar{\succ}_i\}_{i\in\mathbb{N}})$. In this case $\succ_i\to\succ\equiv\gamma\mathbb{R}_+^{2n}\setminus Ls(\{\bar{\succ}_i\}_{i\in\mathbb{N}})$ (for a proof and references cf. W.HILDENBRAND [13] ibidem). Hence we obtained a convergence criterion for (P,δ) based on sequences in $\gamma\mathbb{R}_+^{2n}$.

We hope that the above introduction of the space (P,δ) of preferences was sufficiently detailed and concrete enough to get a first comprehension of this necessary, technical tool. In the following we want to apply (P,δ) to deduce explicitly some more results contained in the first chapter.

PROPOSITION A.7 Let $S \subset P_{mo}$. Let cl(S) be the closure of S in the metric space (P,δ). Then $cl(S) \subset P_{mo}$ iff the following holds: For every pair $(x,y) \in \mathbb{R}_+^n \times \mathbb{R}_+^n$ with $x>y$ there is an open subset $U(x,y)$ of $\mathbb{R}_+^{2n}$ such that $(x,y) \in U(x,y) \subset \succ$ for every $\succ \in S$.

P r o o f: At first we prove the if part: Let $\succ_o$ be an accumulation point of S in P with respect to δ. Let $(x,y) \in \mathbb{R}_+^{2n}$ with $x>y$. By the assumption there is an open subset $U(x,y)$ of $\mathbb{R}_+^{2n}$ such that $(x,y) \in U(x,y) \subset \succ$ for every $\succ \in S$. It follows $(x,y) \in \succ_o$. Therefore $\succ_o$ is monotonic. As $\succ_o$ was chosen arbitrarily in cl(S) we obtain $cl(S) \subset P_{mo}$.

We show the only if part by a contradiction: Let $cl(S) \subset P_{mo}$ and suppose there would be $(x_o,y_o) \in \mathbb{R}_+^n \times \mathbb{R}_+^n$, $x_o>y_o$, such that no open set $U(x_o,y_o)$ would exist in $\mathbb{R}_+^{2n}$ with $U(x_o,y_o) \subset \succ$ for every $\succ \in S$. Then there would be a sequence $\{\succ_i\}_{i\in\mathbb{N}} \subset S$ and a sequence $\{(x_i,y_i)\}_{i\in\mathbb{N}}$ with $(x_i,y_i) \in \mathbb{R}_+^n \times \mathbb{R}_+^n \setminus \succ_i \; \forall i \in \mathbb{N}$ such that $(x_i,y_i) \rightarrow (x_o,y_o)$ for $i \to \infty$. Since the metric space (P,δ) is compact we can assume without loss of generality that $\{\succ_i\}_{i\in\mathbb{N}}$ would be convergent: $\succ_i \rightarrow \succ_o \in cl(S)$. It would follow that $(x_o,y_o) \in \mathbb{R}_+^n \times \mathbb{R}_+^n \setminus \succ_o$. This would be a contradiction, because $\succ_o \in cl(S) \subset P_{mo}$ i.e. $(x_o,y_o) \in \succ_o$ for $x_o>y_o$. Q.E.D.

COROLLARY A.8 Let N be a dense subset of the compact space K. Assume that the mapping $f: N \rightarrow P_{mo}$ possesses a continuous extension $\bar{f}: K \rightarrow P$. Then $\bar{f}(K) \subset P_{mo}$ iff for every pair $(x,y) \in \mathbb{R}_+^n \times \mathbb{R}_+^n$ with $x>y$ there is an open subset $U(x,y)$ of $\mathbb{R}_+^{2n}$ fulfilling $(x,y) \in U(x,y) \subset f(i)$ for every $i \in N$.

Let $\{\alpha(i)\}_{i\in\mathbb{N}}$ be a sequence of scale observations of $\mathbb{R}_+$. Then $\{\alpha(i)\}_{i\in\mathbb{N}}$ is called *growing* if $r(\alpha(i)) \to \infty$ and $grc(\alpha(i)) \to 0$ for $i \to \infty$.

The definition is applied analogously for scale observations of $\mathbb{R}_+^n$.

PROPOSITION A.9 Let $\{\alpha^n(i)\}_{i\in\mathbb{N}}$ be a growing sequence of scale observation of $\mathbb{R}_+^n$. Let $\varepsilon>0$. Then there is an $i_o\in\mathbb{N}$ such that the following holds: For every $\succ\in P_{mo}$ and for every $i\geq i_o$ we have $\delta(\succ,\succ(\alpha^n(i)))<\varepsilon$.

P r o o f: At first we introduce several notations: $\|x\|_\infty:=\max\{|x_i| \mid 1\leq i\leq n\}$ is the max. norm on $\mathbb{R}^n$ and $\|\ \|$ is the Euclidean norm. Let $K_+^{2n}(r):=\{x\in\mathbb{R}_+^{2n} \mid \|x\|_\infty\leq r\}$ the cube in $\mathbb{R}_+^{2n}$ with edge length $r>0$. If $x,y\in\gamma\mathbb{R}_+^{2n}$ then $m(x,y):=\|p^{-1}(x)-p^{-1}(y)\|$ denotes the metric induced by the spherical projection. We denote $\gamma\mathbb{B}_m^{2n}(F,\eta):=\{z\in\gamma\mathbb{R}_+^{2n} \mid m(F,z)<\eta\}$ where $F\subset\gamma\mathbb{R}_+^{2n}$ and $\eta>0$. Analogously $\mathbb{B}_m^{2n}(F,\eta):=\{z\in\mathbb{R}_+^{2n} \mid m(F,z)<\eta\}$ where $F\subset\mathbb{R}_+^{2n}$ and $\eta>0$. Observe that there is a constant $\Omega>1$ with $m(x,y)\leq\Omega\,\|x-y\|$ for $x,y\in\mathbb{R}_+^{2n}$.

Let $\varepsilon>0$. If $i_1\in\mathbb{N}$ is large enough then $\gamma\mathbb{R}_+^{2n}\setminus K_+^{2n}(r(\alpha(i)))\subset\gamma\mathbb{B}_m^{2n}(\{\infty\},\varepsilon$ Hence it suffices to prove that there is an $i_o\in\mathbb{N}$, $i_o\geq i_1$ such that for every $\succ\in P_{mo}$ and for every $i\geq i_o$ the following holds:

(1) $\overline{\succ(\alpha^n(i))}\cap K_+^{2n}(r(\alpha(i_1)))\subset\mathbb{B}_m^{2n}(\overline{\succ}\setminus\{\infty\},\varepsilon)$ and

(2) $\overline{\succ}\cap K_+^{2n}(r(\alpha(i_1)))\subset\mathbb{B}_m^{2n}(\overline{\succ(\alpha^n(i))}\setminus\{\infty\},\varepsilon)$.

We prove that there is an $i_2\geq i_1$ such that (1) is fulfilled for $i\geq i_2$: Let $i_2\in\mathbb{N}$ such that $grc(\alpha(i))<\varepsilon/(4n\Omega)$ for every $i\geq i_2$. Let $\succ\in P_{mo}$. Let $i\geq i_2$ and let $z\in\overline{\succ(\alpha^n(i))}\cap K_+^{2n}(r(\alpha(i_1)))$. Due to the grid construction and the definition of an observed preference there are grid points $s_\xi, s_{\xi'}$ of $\alpha^n(i)$ such that $s=(s_\xi,s_{\xi'})\in\overline{\succ}$ and $\|s-z\|_\infty\leq 2grc(\alpha(i))$. Hence $m(s,z)\leq\Omega\|s-z\|$ $\leq 2n\Omega\,\|s-z\|_\infty\leq 2n\Omega 2grc(\alpha(i))<4n\Omega\varepsilon/(4n\Omega)=\varepsilon$. Therefore $z\in\mathbb{B}_m^{2n}(\overline{\succ}\setminus\{\infty\},\varepsilon)$ and (1) holds for $i\geq i_2$.

We prove that there is an $i_3\geq i_1$ such that (2) is fulfilled for $i\geq i_3$: We choose $i_3\in\mathbb{N}$ such that $grc(\alpha(i))<\varepsilon/(4n\Omega)$ and $r(\alpha(i))>r(\alpha(i_1))+2\varepsilon$ holds for every $i\geq i_3$. Let $\succ\in P_{mo}$ and let $z=(x,y)\in\overline{\succ}\cap K_+^{2n}(r(\alpha(i_1)))$. Then $z_-:=\{(\bar{x},\bar{y})\in\mathbb{R}_+^n\times\mathbb{R}_+^n \mid \bar{x}$ is strictly smaller than x in every component and y is strictly smaller than $\bar{y}$ in every component$\}\subset\overline{\succ}$. Due to the grid construction and the definition of an observed preference the inclusion (2) follows in analogy to (1) for $i \geq i_3$.

Hence with $i_o := \max\{i_2, i_3\}$ the proposition follows. Q.E.D.

A.3.2 A PROOF CONCERNING CHAPTER III. In this section we apply the result of T.E.ARMSTRONG and M.K.RICHTER [2] to prove the Core-Walras equivalence within our framework. Since our approach is a specialized one it suffices mostly to go back to an earlier version of their model, namely to that of M.K.RICHTER [A8]. Naturally within this purely technical section we assume some familiarity with the referred approach.

We assume that $\overline{\mathcal{E}} := \bar{E} \times \bar{e} : (\mathrm{Stone}(R_o), \overline{\mathcal{R}}_o, \bar{d}) \rightarrow P_{mo} \times \mathbb{R}^n_+$ is an observed market with pure competition. Moreover we assume that $\overline{\mathcal{E}}$ possesses a positive first endowment.

By the proof of prop. III.12 it remains to prove that $\mathcal{C}(\overline{\mathcal{E}}) \subset W(\overline{\mathcal{E}})$ holds. To avoid unnecessary notations we restrict ourselves onto the *deterministic* case.

We introduce the following notations: The Boolean algebra of clopen subsets of Stone(R_o) is abbreviated by $\bar{R}_o$. Let $C_+ := \{\bar{f} : \mathrm{Stone}(R_o) \rightarrow \mathbb{R}^n_+ \mid \bar{f}$ is continuous$\}$ and let $C = C_+ - C_+ := \{\bar{f} : \mathrm{Stone}(R_o) \rightarrow \mathbb{R}^n \mid \bar{f}$ is continuous$\}$. If $\bar{f} \in C$ then the finitely additive vector measure $\bar{f}' : \bar{R}_o \rightarrow \mathbb{R}^n$ induced by $\bar{f}$ and $\bar{d}$ is defined by

$$\bar{f}'(\bar{A}) := \int_{\bar{A}} \bar{f} d\bar{d}$$

for every $\bar{A} \in \bar{R}_o$. We will need the following result (for a proof cf. T.E. ARMSTRONG and M.K.RICHTER [2] lemma 4):

LEMMA A.10. Let $\bar{f}_1, \bar{f}_2 \in C$. Let $\varepsilon > 0$, let $t_1, t_2 \in [0,1]$ with $t_1 + t_2 = 1$, and let $\bar{A}_1, \bar{A}_2 \in \bar{R}_o$. Then there exist $\bar{B}_1, \bar{B}_2 \in \bar{R}_o$, $\bar{B}_i \subset \bar{A}_i$ for $i=1,2$, $\bar{B}_1 \cap \bar{B}_2 = \emptyset$ with

$$\| \bar{f}_i'(\bar{B}_i) - t_i \bar{f}_i'(\bar{A}_i) \| < \varepsilon$$

for $i=1,2$.

Assume that $\bar{f}\in \mathcal{C}(\bar{\mathcal{E}})$. We have to demonstrate that $\bar{f}\in W(\bar{\mathcal{E}})$. As usual we apply at first the Hahn-Banach separation theorem:

For $\bar{A}\in\bar{R}_o$, $\bar{A}\neq\emptyset$ we define $G(\bar{A}):=\{\bar{g}'(\bar{A})-\bar{e}'(\bar{A})\,|\,\bar{g}\in C_+ \text{ and } \bar{g}\succ_{\bar{A}}\bar{f}\}$. Let $G:=\cup\{G(\bar{A})\,|\,\bar{A}\in\bar{R}_o \text{ and } \bar{d}(\bar{A})>0\}$. We show that cl(G) is convex:

Let $X_1:=(\bar{g}_1'(\bar{A}_1)-\bar{e}'(\bar{A}_1))\in G(\bar{A}_1)$, $X_2:=(\bar{g}_2'(\bar{A}_2)-\bar{e}'(\bar{A}_2))\in G(\bar{A}_2)$ where $\bar{A}_1$, $\bar{A}_2\in\bar{R}_o$. Let $t\in(0,1)$, let $t_1=t$, $t_2=(1-t)$, and let $X=t_1X_1+t_2X_2$. Assume that $\varepsilon>0$. We prove that there is a $y\in G$ with $\|x-y\|<\varepsilon$:

We denote $\zeta_i:=(\bar{g}_i-\bar{e})'$ for $i=1,2$. By the lemma A.10 there exist $\bar{B}_1$, $\bar{B}_2\in\bar{R}_o$ such that $\bar{B}_i\subset\bar{A}_i$, $i=1,2$, $\bar{B}_1\cap\bar{B}_2=\emptyset$ and $\|\zeta_i(\bar{B}_i)-t_iX_i(\bar{A}_i)\|<\varepsilon/2$ for $i=1,2$.

We define $\bar{g}\equiv\bar{g}_i$ on $\bar{B}_i$, $i=1,2$, and $\bar{g}\equiv\bar{e}$ elsewhere. Then $\bar{g}\in C_+$. If $\bar{B}=\bar{B}_1\cup\bar{B}_2$ then $\bar{g}\succ_{\bar{B}}\bar{f}$. Therefore $\bar{g}'(\bar{B})-\bar{e}'(\bar{B})\in G(\bar{B})\subset G$. Since $\bar{g}'(\bar{B}')-\bar{e}'(\bar{B})=\bar{g}_1'(\bar{B}_1)-\bar{e}'(\bar{B}_1)+\bar{g}_2'(\bar{B}_2)-\bar{e}'(\bar{B}_2)=\zeta_1(\bar{B}_1)+\zeta_2(\bar{B}_2)=:y$ we obtain $\|x-y\|<\varepsilon$.

The convexity of cl(G) is easily deduced now for the Euclidean norm fulfills the triangular inequality.

In the next step we show that $0\notin$ int cl(G): If $0\in$ int cl(G) then there would exist a purely negative $z\in$ cl(G). Therefore there would exist a purely negative $z'\in G$. Hence $\bar{f}$ would be unacceptable with respect to a suitable, non void $\bar{A}\in\bar{R}_o$. This would be a contradiction to the definition of G for $\bar{f}$ is an element of the Core $\mathcal{C}(\bar{\mathcal{E}})$. Hence $0\notin$ int cl(G).

Since $0\notin$ int cl(G) and since cl(G) is convex, separating hyperplane theorems then guarantee the existence of a $p\in\mathbb{R}^n\setminus\{0\}$ weakly separating 0 from cl(G); i.e. $p\cdot x\geq 0$ for all $x\in$ cl(G).

Let $\bar{A}\in\bar{R}_o$ and $\bar{g}\in C_+$ such that $\bar{d}(\bar{A})>0$ and $\bar{g}\succ_{\bar{A}}\bar{f}$. Then $\bar{g}'(\bar{A})-\bar{e}'(\bar{A})\in G(\bar{A})$. Therefore $p\cdot\bar{g}'(\bar{A})\geq p\cdot\bar{e}'(\bar{A})$. Since this inequality holds as well for every $\bar{B}\in\bar{R}_o$, $\bar{B}\subset\bar{A}$, we obtain $p\cdot\bar{g}(x)\geq p\cdot\bar{e}(x)$ for every $x\in\bar{A}$ by the continuity of

$\bar{g}$ and $\bar{e}$. Hence we obtained:

(+) If $\emptyset \neq \bar{A} \in \bar{R}_o$, $\bar{g} \in C_+$, and $\bar{g} \succ_{\bar{A}} \bar{f}$ then $p \cdot \bar{g}(x) \geq p \cdot \bar{e}(x)$ for every $x \in \bar{A}$.

At next we prove $p \cdot \bar{f}(x) \leq p \cdot \bar{e}(x)$ for every $x \in \text{Stone}(R_o)$ as follows: By the monotonicity assumption - i.e. $\bar{E}(\text{Stone}(R_o)) \subset P_{mo}$ - there are $\bar{g} \in C_+$ arbitrarily closed to $\bar{f}$ and preferred to $\bar{f}$ - e.g. take $\bar{g} = \bar{f} + (\varepsilon, \ldots, \varepsilon)$ with $\varepsilon > 0$. Therefore $p \cdot \bar{f}'(\bar{A}) \geq p \cdot \bar{e}'(\bar{A})$ holds for every $\bar{A} \in \bar{R}_o$.

If there is an $\bar{A}_o \in \bar{R}_o$ with $p \cdot \bar{f}'(\bar{A}_o) > p \cdot \bar{e}'(\bar{A})$ then we obtain $p \cdot \bar{f}'(\text{Stone}(R_o)) > p \cdot \bar{e}'(\text{Stone}(R_o))$. This contradicts that $\bar{f}$ is an element of the Core: Indeed, by the monotonicity assumption an element of the Core fulfills $\bar{f}'(\text{Stone}(R_o)) = \bar{e}'(\text{Stone}(R_o))$.

Therefore $p \cdot \bar{f}'(\bar{A}) \leq p \cdot \bar{e}'(\bar{A})$ for every $\bar{A} \in \bar{R}_o$. Since $\bar{f}, \bar{e}$ are continuous we obtain:

(++) $p \cdot \bar{f}(x) \leq p \cdot \bar{e}(x)$ holds for every $x \in \text{Stone}(R_o)$.

It remains to prove that p is strictly positive. We already know that $p \neq 0$. At first we deduce by a contradiction that $p > 0$ - i.e. that $p_i \geq 0$ holds for $1 \leq i \leq n$.

Assume that $p > 0$ does not hold. Without loss of generality we can assume then that $p_1 < 0$. Then we define $\bar{h} \in C_+$ by $\bar{h}'(\bar{A}) := \bar{f}'(\bar{A}) + (\bar{e}'(\bar{A}), 0, 0, \ldots, 0)$ for every $\bar{A} \in R_o$; i.e. $\bar{h} = \bar{f} + (\sum_{i=1}^{n} \bar{e}_i, 0, 0, \ldots, 0)$. By the monotonicity we obtain $\bar{h} \succ_{\text{Stone}(R_o)} \bar{f}$. Since $\bar{f}'(\text{Stone}(R_o)) = \bar{e}'(\text{Stone}(R_o))$ we obtain $p \cdot \bar{h}'(\text{Stone}(R_o)) < p \cdot \bar{e}'(\text{Stone}(R_o))$. This contradicts the assumption that p weakly separates 0 from cl(G). Hence $p > 0$ follows.

In the next step we prove that p is strictly positive; i.e. that $p_i > 0$ for $1 \leq i \leq n$. As before this is shown by a contradiction:

Without loss of generality we can assume $p_1 = 0$ and $p_2 > 0$. By the preceding paragraphs we know $\bar{h} \succ_{\text{Stone}(R_o)} \bar{f}$. Due to our definition of the coalitional-preference there exists an $\varepsilon > 0$ and a $\bar{g} \in C_+$ such that $\bar{g}'(\text{Stone}(R_o)) = \bar{h}'(\text{Stone}(R_o)) - (0, \varepsilon, 0, 0, \ldots, 0)$ and $\bar{g} \succ_{\text{Stone}(R_o)} \bar{f}$ holds. By assumption $\bar{e}'(\text{Stone}(R_o))$ is strictly positive. Moreover $\bar{h}'(\text{Stone}(R_o)) >$

$\bar{f}'(\text{Stone}(R_o))=\bar{e}'(\text{Stone}(R_o))$. We obtain

$$
\begin{aligned}
&p\cdot\bar{e}'(\text{Stone}(R_o))\leq p\cdot\bar{g}'(\text{Stone}(R_o))\\
&=p\cdot\bar{h}'(\text{Stone}(R_o))-p_2\varepsilon\\
&=p\cdot\bar{f}'(\text{Stone}(R_o))-p_2\varepsilon\\
&=p\cdot\bar{e}'(\text{Stone}(R_o))-p_2\varepsilon\\
&<p\cdot\bar{e}'(\text{Stone}(R_o)).
\end{aligned}
$$

Hence $p\cdot\bar{e}'(\text{Stone}(R_o))<p\cdot\bar{e}'(\text{Stone}(R_o))$ which is a contradiction. Therefore p is strictly positive.

By (+), (++), and $p\in\mathbb{R}^n_{++}$ it follows that $\bar{f}\in W(\bar{\mathcal{E}})$ due to our definition of a p-Walras allocation. Q.E.D.

A4 THE ELEMENTARY MODEL

Within this fourth part of the appendix we collected all definitions and results concerning the elementary representation of the model. A comment on the whole model concerning its relevance and its degree of technical adjustment to the original economic problem is added.

A.4.1 THE ELEMENTARY REPRESENTATION. Here we want to present all material concerning the elementary representation of the model. However we are unable to recall motivations here as well. For those we refer the reader to the first three chapters, in particular to the verbal explanations.

The mathematical background needed here is the classical Riemann integration theory and the compact metric space (P,δ) of preferences. In the case that the reader is not informed about all aspects of B.Riemann's integration approach he is referred to the part A2 of this appendix. An introduction of the space (P,δ) is contained in part A.3.1.

At first we introduce some notations. Let (M,m) be one of the following metric spaces (P,δ), $(\mathbb{R}^n_+, \| \ \|)$, or $(P \times \mathbb{R}^n_+, \delta \times \| \ \|)$ where $\| \ \|$ denotes the Euclidean distance. Let f be a bounded, M-valued function defined on the unit interval [0,1]. Then f is called *right continuous* iff $\lim_{x \downarrow a} f(x) = f(a)$ holds for every $a \in [0,1)$. Here $x \downarrow a$ means $x \to a$ and $x > a$. The function f is said to be *left continuously extensible* iff $\lim_{x \uparrow a} f(x)$ exists for every $a \in (0,1]$. Here $x \uparrow a$ means $x \to a$ and $x < a$. If $0 \leq a < b \leq 1$ then the characteristic function $\chi_{[a,b)}$ of the half open interval [a,b) is right continuous, left continuously extensible. Remember that $\chi_{[a,b)}(x) = 1$ if $x \in [a,b)$ and that $\chi_{[a,b)}(x) = 0$ if $x \in [0,1]$, $x \notin [a,b)$.

We denote $REP([0,1],M) := \{f:[0,1] \to M \mid f$ is bounded, right continuous, left continuously extensible, and continuous in every $x \in [0,1] \setminus ((0,1) \cap \mathbb{Q})\}$. On REP([0,1],M) we consider the sup metric m_0 defined by $m_0(f,g) := \sup\{m(f(x),g(x)) \mid x \in [0,1]\}$. One can prove that $(REP([0,1],M), m_0)$ is a complete metric space; i.e. every Cauchy sequence is convergent.

The behaviour of the space REP([0,1],M) is widely the same as that of the space of continuous functions on [0,1]. Indeed the 'REP-functions' can be seen as 'broken-continuous' functions. Observe that $REP([0,1],\mathbb{R}_+)$ contains all characteristic functions $\chi_{[a,b)}$ where $0 \leq a < b < 1$ and $a,b \in \mathbb{Q}$.

The idealized status of pure competition holds in a exchange market model if the influence of every single member onto the market situation is zero. Hence the market situation remains unchanged if the preferences or the commodity bundles of one or finitely many members of the market were arbitrarily altered. Therefore a market with pure competition corresponds in a natural way to an equivalence class of preference-endowment functions. The technical step allowing us to handle such a viewpoint is performed as follows:

Let $\mathcal{D}([0,1],\lambda,M) := \{f:[0,1] \to M \mid f$ is continuous except on a subset A_f of [0,1] having Lebesgue measure zero$\}$. The essential sup metric m_{ess} on $\mathcal{D}([0,1],\lambda,M)$ is defined by $m_{ess}(f,g) := \inf\{c > 0 \mid m(f(x),g(x)) < c$ for every $x \in [0,1]$ except on a Lebesgue zero set$\}$. Identifying in

$\mathcal{D}([0,1],\lambda,M)$ all functions having an m_{ess}-zero-distance yields the space $D([0,1],\lambda,M)$ consisting of equivalence classes of functions. Then m_{ess} is a metric on $D([0,1],\lambda,M)$ and $(D([0,1],\lambda,M),m_{ess})$ is complete.

We denote $REP([0,1],\lambda,M)$ the subspace of $D([0,1],\lambda,M)$ which is generated by $REP([0,1],M)$; i.e. $REP([0,1],\lambda,M)$ consists of those equivalence classes in $D([0,1],\lambda,M)$ which contain an element of $REP([0,1],M)$. Observe that two different elements of $REP([0,1],M)$ are never in the same equivalence class. One can prove that $(REP([0,1],\lambda,M),m_{ess})$ is a complete metric space.

In the case that the reader dislikes an explicit calculation by equivalence classes he should not be worried. In the following one can always use $(REP([0,1],M),m_o)$ instead of $(REP([0,1],\lambda,M),m_{ess})$ without affecting the results. However concerning the interpretation one should keep in mind that sufficiently small disturbances do not alter the market situation.

A sequence $\{x_i\}_{i\in \mathbb{N}} \subset [0,1]$ is *uniformly distributed* if $\lim_{n\to\infty} (1/n)\Sigma_{i=1}^{n} f(x_i) = \int_{[0,1]} f(t)dt$ holds for every Riemann integrable function f. Concerning $\lambda \times \lambda \times \lambda \times \ldots$ on $[0,1] \times [0,1] \times [0,1] \times \ldots$ almost all $(x_1,x_2,x_3,\ldots)$ are uniformly distributed, i.e. such an assumption corresponds to the law of large numbers.

We are sufficiently prepared now to introduce the represented model. For the sake of simplicity and, as we hope, clarity we restrict ourselves at first entirely onto the *represented observed deterministic model*. Later on we will show in one final step how the other cases can be deduced. For convenience we will use here a *notation slightly different* from the previous one, i.e. we will drop all bars $^{-}$.

Let $[0,1]$ be the unit interval. The rational points contained in $[0,1)$ are called represented *agents*. All other points of the unit interval are corresponding to non realized property combinations.

Half open intervals $[a,b) \subset [0,1]$ with rational boundary points a,b as well as finite unions of such intervals are called represented observed *coalitions*. The *natural influence* of a represented observed coalition is given by the Jordan content; i.e. the natural influence of $[a,b) \subset [0,1]$ is b-a.

A mapping $f:[0,1] \rightarrow \mathbb{R}^n_+$ with $f \in REP([0,1],\mathbb{R}^n_+)$ is called a represented deterministic observed *allocation*. A mapping $E:[0,1] \rightarrow P_{mo}$ with $E \in REP([0,1],P)$ is called a represented deterministic observed *profile*. Such a profile is *monotonic* provided that the closure of the set $E([0,1])$ is contained in P_{mo}; i.e. iff $cl(E([0,1])) \subset P_{mo}$ holds. Here P_{mo} denotes the set of monotonic preferences.

The above closure condition is a technical trick to exclude cases which are possible by mathematics but irrelevant within our microeconomic model. The condition is fulfilled iff $\lim_{x \uparrow a} E(x) \in P_{mo}$ holds for every rational point in the unit interval. An other criterion to check the condition is prop. A.7.

Let e be an allocation and let E be a monotonic profile. Then $\mathcal{E} := E \times e:[0,1] \rightarrow P_{mo} \times \mathbb{R}^n_+$ is called a represented deterministic observed *market mapping*. A represented deterministic observed *market* with pure competition is symbolized by $\mathcal{E} := E \times e:([0,1]$, the class of coalitions, the natural influence$) \rightarrow P_{mo} \times \mathbb{R}^n_+$ where $\mathcal{E}$ is a market mapping. For convenience we will use in the following the abbreviations $\mathcal{E}$ resp. $\mathcal{E} = E \times e$ to symbolize a market.

A *positive first endowment* prevails in a market $\mathcal{E} = E \times e$ iff the endowment per capita is positive with respect to every commodity, i.e.

$$\int_{[0,1]} e_i(t)\,dt > 0 \text{ for } 1 \leq i \leq n,$$

and with respect to every non void coalition; i.e. there is an $\varepsilon > 0$ with

$$\max\{e_i(t) \mid 1 \leq i \leq n\} > \varepsilon$$

for every $t \in [0,1]$.

Remark that by J.v.Neumann's rearrangement theorem there is an enumeration $\{x_i\}_{i\in\mathbb{N}}$ of the rational numbers contained in $[0,1)$ - i.e. of the agents - such that

$$\lim_{n\to\infty} \frac{1}{n}\sum_{j=1}^{n} e_i(x_j)=\int_{[0,1]} e_i(t)\,dt$$

holds.

In the following we always *assume* that $\mathcal{E}=E\times e$ is a market with a positive first endowment.

An allocation f is called *attainable* iff

$$\int_{[0,1]} f_i(t)\,dt \leq \int_{[0,1]} e_i(t)\,dt$$

holds for $1\leq i\leq n$.

Let $\succ\in P_{mo}$ be a monotone preference. Let $x,y\in\mathbb{R}_+^n$ and let $\varepsilon>0$. Then x is called ε-*preferred* to y by $\succ$ iff $U_\varepsilon((x,y)):=\{z\in\mathbb{R}_+^{2n} \mid \| z-(x,y)\| <\varepsilon\}\subset\succ$. Let f,g are two allocations and let A be a non void coalition. Then f is called ε-*preferred to* g *by* A - denoted: f ε-$\succ_A$g - iff f(t) is ε-preferred to g(t) by E(t) for every $t\in A$. If there exists an $\varepsilon>0$ such that f is ε-preferred to g by A then f is *preferred to* g *by* A - denoted: $f \succ_A g$.

We call f A-*unacceptable by* g if (a) $\int_A g_i(t)\,dt<\int_A e_i(t)\,dt \quad \forall$ i but $=$ if $\int_A e_i(t)\,dt=0$ and (b) $g \succ_A f$ are fulfilled. Then f is A-*unacceptable* if there is an allocation g such that f is A-unacceptable by g. Lastly f is *unacceptable* if there is a non void allocation A such that f is A-unacceptable.

The *Core* $\mathcal{C}(\mathcal{E})$ of $\mathcal{E}$ consists of all attainable, non unacceptable allocations.

We denote $\mathbb{R}_{++}^n:=\{x\in\mathbb{R}^n \mid x_i>0 \text{ for } 1\leq i\leq n\}$. If $p\in\mathbb{R}_{++}^n$ then p is called a *price vector*.

Let p be a price vector. Let $\varepsilon>0$. Then f is called ε-p-*attainable* if $(\sum_{i=1}^n p_i f_i(t)+\varepsilon<\sum_{i=1}^n p_i e_i(t)$ is fulfilled for every $t\in[0,1]$. We call f *positive*-p-*attainable* if there is an $\varepsilon>0$ such that f is ε-p-attainable.

Lastly f is p-*attainable* if f is the limit (with respect to the sup. norm) of a sequence of positive p-attainable allocations.

A p-attainable allocation f is called p-*unacceptable* if there is a positive p-attainable allocation g and a non void coalition A such that $g \succ_A f$.

An attainable, p-attainable, and not p-unacceptable allocation is called a p-*Walras allocation*. An allocation f is a *Walras allocation* if there is a price vector p such that f is a p-Walras allocation. We denote $W(\mathcal{E})$ the set of all Walras allocations of $\mathcal{E}$.

Let $\mathcal{E} := E \times e$ be a market with a positive first endowment; i.e. $E: [0,1] \to P_{mo}$ is a monotonic profile and $e:[0,1] \to \mathbb{R}_+^n$ fulfills the requirements of a positive first endowment. Then $W(\mathcal{E}) = \mathcal{C}(\mathcal{E})$.

By restricting the coalitions and functions used above onto the rational numbers we obtain the represented deterministic model. Naturally, integrals have to be calculated then by J.v.Neumann's theorem.

Interpreting the above model within the $REP([0,1],\lambda,P\times \mathbb{R}_+^n)$ case yields the represented observed probabilistic model: Coalitions are variated then by Jordan zero sets i.e. subsets $A \subset [0,1]$ with $\lambda(cl(A))=0$. Functions are seen as equivalence classes concerning the ess.sup.metric which does not bother results of integrals.

The restriction onto the rational numbers then yields analogously the represented probabilistic case.

A.4.2 A FINAL COMMENT. We want to throw here some light on several aspects of our model which we have not discussed so far.

Compared with the condensed model described above the first three chapters contain a bulk of mathematical technics. One may suspect that a combination of the above section with a concentrate - sufficiently detailed - of the verbal explanations in chapters one to three allow a still satisfactory presentation of the economic problem, at least with respect to practical purposes and applications. Since our personal viewpoint is not absolutely different we want to explain why we described the whole way of the development of our model.

Starting from microeconomic methods of measurement we deduced an abstract deterministic model. Its concretization, its translation into a probabilistic model, and its transformation into an elementary representation was following. Hence, based on the finitely additive measure theory, we constructed a specialized probability model which allowed a precise description of the interaction with the underlying deterministic model. All that was in our opinion not only a contribution to microeconomics but also a contribution to stochastics, in particular to the foundations of probability theory. Naturally it caused a great investment into the required mathematics.

Provided that the specialized probability model developed here could be borrowed from a well established textbook in mathematics then we agree that our research report could be condensed strongly. Unfortunately we don't know such a monograph. A lot of the required technics we had to develop on our own.

When we considered in our context the finitely additive measure theory as a specialization of the σ-additive one then we have to report the other side of the shield also. Naturally every σ-additive measure is a specialized finitely additive measure. This further aspect was succesfully applied in the fine model of T.E.ARMSTRONG and M.K.RICHTER [2], [3].

We explained above why our model needed a lot of mathematical efforts. On the other hand, in our opinion the model needs even more investments in mathematical tools to improve its adjustment to microeconomic requirements:

We described the microeconomic problem considered here by continuous resp. broken continuous functions. In sciences dealing with reality one usually tries to describe the considered situation by differentiable functions and differential equations. Hence it is natural to probe a simplification of our model by a reduction onto differentiable functions resp. broken differentiable functions.

Faced with such a goal, at first a concept of a space of preferences is necessary, different to the (P,δ) used here. It should be based and motivated strictly by microeconomic methods of measurement. Naturally, it has to include a completion concept. We have some doubts whether the application of available mathematical tools - mostly based on earlier developments in physics - can fulfill these requirements immediatly.

A rigorous grounding of such a differential approach on microeconomic methods of measurement - analogous to chapter one - requires the development of a differential approach on compact supports of Radon measures. It may be that specialized rearrangement results and motions are helpful for this purpose. Without any doubt the confrontation with a problem of that quality is discouraging.

The only spark of hope seems to us to consists of two possible concepts of solution; one concerning the concretized case - i.e. Stone(R_o) resp. $w([0,1])$ - and the other one concerning the model represented elementarily:

A verbal formulation of the meaning of differentiability once heard by me in physics reads as follows: A real valued function on $\mathbb{R}$ is differentiable provided that it can be described locally in first approximation by proportional functions i.e. linear functions. On $w([0,1])$ the meaning of the word 'locally' is given by the metric. The linear functions on $[0,1]$ extend uniquely to a class L' of functions on $w([0,1])$. Applying the above definition for L', we obtain a differential calculus on $w([0,1])$ resp. on Stone(R_o). Within such a calculus the characteristic functions of coalitions turn out to be differentiable infinitely many times. For our purposes this behaviour is quite natural.

The idea concerning the model represented elementarily is the following: Instead of 'broken-continuous' functions on [0,1] we can use 'broken-differentiable' functions. A general, abstract concept to handle such problems can be found in Z.SEMADENI [33] § 15. Since characteristic functions of represented coalitions turn out to be 'broken-differentiable', the completion concept of chapter two can be applied analogously.

We conclude that the technical adjustment of our model to requirements of microeconomic research probably can be still perfected.

It may be that the reduction indicated above of our model onto generalized differentiable mappings can be realized. If this would be true then generalized differential equations perhaps could allow a handling of the dynamical situation - that is at least our conjecture. Probably the definitions of equilibria elaborated in the third chapter then must be reformulated according to stability criteria for solution paths of (generalized) differential equations.

A third aspect of our model is its strong specialization. We studied an exchange market only. No production and no other topics are considered by us. However, at least in other sciences concerning reality it is a fruitful tradition to describe and scrutinize at first specialized, typical situations before more complicated problems are investigated.

We scetched above that our model is not perfectly condensed, probably still not perfectly developed concerning its technical handling, and rather specialized. It is our natural desire to illustrate some of its advantages also:

The system of available methods of measurement is of an essential importance within every science dealing with reality. Typical methods of measurement used in microeconomics - i.e. public opinion polls and microcensuses - possess a qualitative advantage compared with macroeconomic measurements: They are quicker and theyare cheaper - provided that single ones are applied only. Therefore they grant the discipline of microeconomics a value of its own. Consequently, microeconomic models should be grounded on microeconomic measuring. Our model fulfills this requirement.

An advantage of theoretical, microeconomic models is their ability to explain economic interrelations resulting in the behaviour and the endowment of households or individuals. A natural assumption for this ability is the possibility of a suitable interpretation of the considered model. In particular strongly idealized models require a precise interpretation. Our model is interpreted in detail - at least so far as the status of pure competition is involved.

We conclude that the developement of our model can not be considered to be finished. Advantages of it are its grounding on microeconomic methods of measurement, its clear interpretability, and its already rather easy handling. We hope that our model can be seen as a seed for microeconomic models of pure competition which are well to interpret, elegant to handle, and which allow the treatment not only of the static, but also of the comparativ static and the dynamic situation.

REFERENCES

1.) REFERENCES CONCERNING CHAPTER I - III

[1] ARMSTRONG, T.E.: Remarks Related to Finitely Additive Exchange Economies, Proceedings Indianapolis 1984, Lecture Notes in Economics and Math. Systems 244 (1985).

[2] ARMSTRONG, T.E. and RICHTER, M.K.: The Core Walras Equivalence, Journal of Economic Theory 33 (1984), 116 - 151.

[3] ARMSTRONG, T.E. and RICHTER, M.K.: Existence of Nonatomic Core-Walras Allocations, Journal of Economic Theory 38 (1986), 137 - 159.

[4] ARMSTRONG, T.E.: Stone Spaces of Algebras of Arithmetic Sequences, Research Report, University of Minnesota 1977.

[5] AUMANN, R.J.: Markets with a Continuum of Traders, Econometrica 32 (1964), 39 - 50.

[6] AUMANN, R.J. and SHAPLEY, L.S.: Values of Non-Atomic Games, Princeton 1974.

[7] BRUNK, H.D.: Stone Spaces in Probability, Oregon University, Dept. of Statistics, Technical Report No. 52 (1976).

[8] DEBREU, G.: On a Theorem of Scarf, Review of Economic Studies 30 (1963), 177 - 180.

[9] DEBREU, G.: Theory of Value, New York 1959.

[10] EDGEWORTH, F.Y.: Mathematical Psychics, London 1881.

[11] HALMOS, P.R.: Measure Theory, New York, Toronto, Melbourne 1973.

[12] HARDY, G.H. and WRIGHT, E.M.: An Introduction to the Theory of Numbers, Oxford 1960.

[13] HILDENBRAND, W.: Core and Equilibria of a Large Economy, Princeton 1974.

[14] HLAWKA, E.: Theorie der Gleichverteilung, Mannheim, Wien, Zürich 1979.

[15] KLEIN, C. and ROLEWICZ, S.: On Riemann Integration of Functions

with Values in Topological Linear Spaces, Studia Math. 80 (1984), 109 - 118.

[16] KLEIN, C.: Invariance Properties of the Banach Algebra of Darboux Integrable Functions, Proc. Karlsruhe 1983, ed. G.Hammer and D.Pallaschke, Lecture Notes in Economics and Math. Systems 226 (1984), 382 - 411.

[17] -, Arzela-Ascoli's Theorem for Riemann Integrable Functions on Compact Spaces, Manuscripta Math. 55 (1986), 403 - 418.

[18] -, On Functions whose Improper Riemann Integral is Absolutely Convergent, Studia Math. 85 (1987), 247 - 255.

[19] -, Atomless Economies with Countably Many Agents, Proc. of the Seminar on Game Theory, Bonn-Hagen 1978, ed. O.Moeschlin and D.Pallaschke, Amsterdam 1979.

[20] -, Closed Subsets of the Set of Monotone Preferences, Methods of Operations Research 41 (1981), 301 - 304.

[21] -, The Observation of a Deterministic Microeconomic Model of a Large Economy, Proc. of the Seminar on Game Theory, Bonn - Hagen 1980, ed. O.Moeschlin and D.Pallaschke, Amsterdam 1981 217 - 231.

[22] -, Finitely Additive Measures on Subsets of Compact Spaces, Meth. of Operations Research 46 (1983), 297 - 305.

[23] KRICKEBERG, K.: Strong Mixing Properties of Markov Chains with Infinite Invariant Measure, Proc. Fifth Berkeley Symp. Math. Stat. Probability 1965 Vol. II, Part 2, 431 - 466, Berkeley and Los Angeles, University of California Press, 1967.

[24] KUIPERS, L. and NIEDERREITER, H.: Uniform Distribution of Sequences, New York, London, Sydney 1974.

[25] KURATOWSKY, K.: Topology I, New York 1966.

[26] LOS, J.: On the Axiomatic Treatment of Probability, Coll. Math. 3 (1955), 125 - 137.

[27] MAS COLELL, A.: The Theory of General Economic Equilibrium, Cambridge 1985.

[28] MORISHIMA, M.: Walras' Economics, Cambridge 1978.

[29] NOELLE-NEUMANN, E.: Umfragen in der Massengesellschaft, Reinbeck 1963.

[30] PALLASCHKE, D.: Markets with Countably Many Traders, Journal of

Applied Math. and Comp. 4 (1978), 201 - 212.

[31] ROLEWICZ, S.: Metric Linear Spaces, Warszawa 1972 and 1984, Dordrecht, Boston, Lancaster 1984.

[32] SCARF, H.: An Analysis of Markets with a Large Number of Participants, Recent Advances in Game Theory, The Princeton University Conference, 1962, 127 - 155.

[33] SEMADENI, Z.: Banach Spaces of Continuous Functions I, Warszawa 1971.

[34] SEMADENI, Z.: Describing q-Adic Solenoids as Maximal Ideal Spaces of Subalgebras of C(R), Functiones et Approximatio XVII, Adam Mickiewicz University Press, Poznan 1987, 97 - 106.

[35] SIKORSKI, R.: Boolean Algebras, Berlin, Heidelberg, New York 1969.

[36] VLADIMIROV, D.A.: Bool'sche Algebren, Berlin 1972.

[37] WALRAS, L.: Elements d'economie politique pure, Lausanne 1874.

2.) REFERENCES CONCERNING THE APPENDIX

[A1] DUNFORD, N. and SCHWARTZ, J.T.: Linear Operators I, New York 1957.

[A2] ENGELKING, R.: General Topology, Warszawa 1977.

[A3] HALMOS, P.R.: Lectures on Boolean Algebras, New York, Heidelberg, Berlin 1974.

[A4] JOHNSON, K.G.: Algebras of Riemann Integrable Functions, Proc. AMS 13 (1962), 437 - 441.

[A5] KELLEY, J.L.: General Toplogy, New York, Heidelberg, Berlin 1955.

[A6] KLINGENBERG, W.: Rimannian Geometry, Berlin, New York 1982.

[A7] NEUMANN, J. von : J.Mat.Fiz.Lapok 32 (1925), 32 - 40.

[A8] RICHTER, M.K.: Coalitions, Core, and Competition, Journal of Economic Theory 3 (1971), 323 - 334.

SUBJECT INDEX

Vol. 211: P. van den Heuvel, The Stability of a Macroeconomic System with Quantity Constraints. VII, 169 pages. 1983.

Vol. 212: R. Sato and T. Nôno, Invariance Principles and the Structure of Technology. V, 94 pages. 1983.

Vol. 213: Aspiration Levels in Bargaining and Economic Decision Making. Proceedings, 1982. Edited by R. Tietz. VIII, 406 pages. 1983.

Vol. 214: M. Faber, H. Niemes und G. Stephan, Entropie, Umweltschutz und Rohstoffverbrauch. IX, 181 Seiten. 1983.

Vol. 215: Semi-Infinite Programming and Applications. Proceedings, 1981. Edited by A.V. Fiacco and K.O. Kortanek. XI, 322 pages. 1983.

Vol. 216: H.H. Müller, Fiscal Policies in a General Equilibrium Model with Persistent Unemployment. VI, 92 pages. 1983.

Vol. 217: Ch. Grootaert, The Relation Between Final Demand and Income Distribution. XIV, 105 pages. 1983.

Vol. 218: P. van Loon, A Dynamic Theory of the Firm: Production, Finance and Investment. VII, 191 pages. 1983.

Vol. 219: E. van Damme, Refinements of the Nash Equilibrium Concept. VI, 151 pages. 1983.

Vol. 220: M. Aoki, Notes on Economic Time Series Analysis: System Theoretic Perspectives. IX, 249 pages. 1983.

Vol. 221: S. Nakamura, An Inter-Industry Translog Model of Prices and Technical Change for the West German Economy. XIV, 290 pages. 1984.

Vol. 222: P. Meier, Energy Systems Analysis for Developing Countries. VI, 344 pages. 1984.

Vol. 223: W. Trockel, Market Demand. VIII, 205 pages. 1984.

Vol. 224: M. Kiy, Ein disaggregiertes Prognosesystem für die Bundesrepublik Deutschland. XVIII, 276 Seiten. 1984.

Vol. 225: T.R. von Ungern-Sternberg, Zur Analyse von Märkten mit unvollständiger Nachfragerinformation. IX, 125 Seiten. 1984

Vol. 226: Selected Topics in Operations Research and Mathematical Economics. Proceedings, 1983. Edited by G. Hammer and D. Pallaschke. IX, 478 pages. 1984.

Vol. 227: Risk and Capital. Proceedings, 1983. Edited by G. Bamberg and K. Spremann. VII, 306 pages. 1984.

Vol. 228: Nonlinear Models of Fluctuating Growth. Proceedings, 1983. Edited by R.M. Goodwin, M. Krüger and A. Vercelli. XVII, 277 pages. 1984.

Vol. 229: Interactive Decision Analysis. Proceedings, 1983. Edited by M. Grauer and A.P. Wierzbicki. VIII, 269 pages. 1984.

Vol. 230: Macro-Economic Planning with Conflicting Goals. Proceedings, 1982. Edited by M. Despontin, P. Nijkamp and J. Spronk. VI, 297 pages. 1984.

Vol. 231: G.F. Newell, The M/M/∞ Service System with Ranked Servers in Heavy Traffic. XI, 126 pages. 1984.

Vol. 232: L. Bauwens, Bayesian Full Information Analysis of Simultaneous Equation Models Using Integration by Monte Carlo. VI, 114 pages. 1984.

Vol. 233: G. Wagenhals, The World Copper Market. XI, 190 pages. 1984.

Vol. 234: B.C. Eaves, A Course in Triangulations for Solving Equations with Deformations. III, 302 pages. 1984.

Vol. 235: Stochastic Models in Reliability Theory. Proceedings, 1984. Edited by S. Osaki and Y. Hatoyama. VII, 212 pages. 1984.

Vol. 236: G. Gandolfo, P.C. Padoan, A Disequilibrium Model of Real and Financial Accumulation in an Open Economy. VI, 172 pages. 1984.

Vol. 237: Misspecification Analysis. Proceedings, 1983. Edited by T.K. Dijkstra. V, 129 pages. 1984.

Vol. 238: W. Domschke, A. Drexl, Location and Layout Planning. IV, 134 pages. 1985.

Vol. 239: Microeconomic Models of Housing Markets. Edited by K. Stahl. VII, 197 pages. 1985.

Vol. 240: Contributions to Operations Research. Proceedings, 1984. Edited by K. Neumann and D. Pallaschke. V, 190 pages. 1985.

Vol. 241: U. Wittmann, Das Konzept rationaler Preiserwartungen. XI, 310 Seiten. 1985.

Vol. 242: Decision Making with Multiple Objectives. Proceedings, 1984. Edited by Y.Y. Haimes and V. Chankong. XI, 571 pages. 1985.

Vol. 243: Integer Programming and Related Areas. A Classified Bibliography 1981–1984. Edited by R. von Randow. XX, 386 pages. 1985.

Vol. 244: Advances in Equilibrium Theory. Proceedings, 1984. Edited by C.D. Aliprantis, O. Burkinshaw and N.J. Rothman. II, 235 pages. 1985.

Vol. 245: J.E.M. Wilhelm, Arbitrage Theory. VII, 114 pages. 1985.

Vol. 246: P.W. Otter, Dynamic Feature Space Modelling, Filtering and Self-Tuning Control of Stochastic Systems. XIV, 177 pages. 1985.

Vol. 247: Optimization and Discrete Choice in Urban Systems. Proceedings, 1983. Edited by B.G. Hutchinson, P. Nijkamp and M. Batty. VI, 371 pages. 1985.

Vol. 248: Plural Rationality and Interactive Decision Processes. Proceedings, 1984. Edited by M. Grauer, M. Thompson and A.P. Wierzbicki. VI, 354 pages. 1985.

Vol. 249: Spatial Price Equilibrium: Advances in Theory, Computation and Application. Proceedings, 1984. Edited by P. T. Harker. VII, 277 pages. 1985.

Vol. 250: M. Roubens, Ph. Vincke, Preference Modelling. VIII, 94 pages. 1985.

Vol. 251: Input-Output Modeling. Proceedings, 1984. Edited by A. Smyshlyaev. VI, 261 pages. 1985.

Vol. 252: A. Birolini, On the Use of Stochastic Processes in Modeling Reliability Problems. VI, 105 pages. 1985.

Vol. 253: C. Withagen, Economic Theory and International Trade in Natural Exhaustible Resources. VI, 172 pages. 1985.

Vol. 254: S. Müller, Arbitrage Pricing of Contingent Claims. VIII, 151 pages. 1985.

Vol. 255: Nondifferentiable Optimization: Motivations and Applications. Proceedings, 1984. Edited by V.F. Demyanov and D. Pallaschke. VI, 350 pages. 1985.

Vol. 256: Convexity and Duality in Optimization. Proceedings, 1984. Edited by J. Ponstein. V, 142 pages. 1985.

Vol. 257: Dynamics of Macrosystems. Proceedings, 1984. Edited by J.-P. Aubin, D. Saari and K. Sigmund. VI, 280 pages. 1985.

Vol. 258: H. Funke, Eine allgemeine Theorie der Polypol- und Oligopolpreisbildung. III, 237 pages. 1985.

Vol. 259: Infinite Programming. Proceedings, 1984. Edited by E.J. Anderson and A.B. Philpott. XIV, 244 pages. 1985.

Vol. 260: H.-J. Kruse, Degeneracy Graphs and the Neighbourhood Problem. VIII, 128 pages. 1986.

Vol. 261: Th.R. Gulledge, Jr., N.K. Womer, The Economics of Made-to-Order Production. VI, 134 pages. 1986.

Vol. 262: H.U. Buhl, A Neo-Classical Theory of Distribution and Wealth. V, 146 pages. 1986.

Vol. 263: M. Schäfer, Resource Extraction and Market Structure. XI, 154 pages. 1986.

Vol. 264: Models of Economic Dynamics. Proceedings, 1983. Edited by H.F. Sonnenschein. VII, 212 pages. 1986.

Vol. 265: Dynamic Games and Applications in Economics. Edited by T. Başar. IX, 288 pages. 1986.

Vol. 266: Multi-Stage Production Planning and Inventory Control. Edited by S. Axsäter, Ch. Schneeweiss and E. Silver. V, 264 pages. 1986.

Vol. 267: R. Bemelmans, The Capacity Aspect of Inventories. IX, 165 pages. 1986.

Vol. 268: V. Firchau, Information Evaluation in Capital Markets. VII, 103 pages. 1986.

Vol. 269: A. Borglin, H. Keiding, Optimality in Infinite Horizon Economies. VI, 180 pages. 1986.

Vol. 270: Technological Change, Employment and Spatial Dynamics. Proceedings 1985. Edited by P. Nijkamp. VII, 466 pages. 1986.

Vol. 271: C. Hildreth, The Cowles Commission in Chicago, 1939–1955. V, 176 pages. 1986.

Vol. 272: G. Clemenz, Credit Markets with Asymmetric Information. VIII, 212 pages. 1986.

Vol. 273: Large-Scale Modelling and Interactive Decision Analysis. Proceedings, 1985. Edited by G. Fandel, M. Grauer, A. Kurzhanski and A.P. Wierzbicki. VII, 363 pages. 1986.

Vol. 274: W.K. Klein Haneveld, Duality in Stochastic Linear and Dynamic Programming. VII, 295 pages. 1986.

Vol. 275: Competition, Instability, and Nonlinear Cycles. Proceedings, 1985. Edited by W. Semmler. XII, 340 pages. 1986.

Vol. 276: M.R. Baye, D.A. Black, Consumer Behavior, Cost of Living Measures, and the Income Tax. VII, 119 pages. 1986.

Vol. 277: Studies in Austrian Capital Theory, Investment and Time. Edited by M. Faber. VI, 317 pages. 1986.

Vol. 278: W.E. Diewert, The Measurement of the Economic Benefits of Infrastructure Services. V, 202 pages. 1986.

Vol. 279: H.-J. Büttler, G. Frei and B. Schips, Estimation of Disequilibrium Models. VI, 114 pages. 1986.

Vol. 280: H.T. Lau, Combinatorial Heuristic Algorithms with FORTRAN. VII, 126 pages. 1986.

Vol. 281: Ch.-L. Hwang, M.-J. Lin, Group Decision Making under Multiple Criteria. XI, 400 pages. 1987.

Vol. 282: K. Schittkowski, More Test Examples for Nonlinear Programming Codes. V, 261 pages. 1987.

Vol. 283: G. Gabisch, H.-W. Lorenz, Business Cycle Theory. VII, 229 pages. 1987.

Vol. 284: H. Lütkepohl, Forecasting Aggregated Vector ARMA Processes. X, 323 pages. 1987.

Vol. 285: Toward Interactive and Intelligent Decision Support Systems. Volume 1. Proceedings, 1986. Edited by Y. Sawaragi, K. Inoue and H. Nakayama. XII, 445 pages. 1987.

Vol. 286: Toward Interactive and Intelligent Decision Support Systems. Volume 2. Proceedings, 1986. Edited by Y. Sawaragi, K. Inoue and H. Nakayama. XII, 450 pages. 1987.

Vol. 287: Dynamical Systems. Proceedings, 1985. Edited by A.B. Kurzhanski and K. Sigmund. VI, 215 pages. 1987.

Vol. 288: G.D. Rudebusch, The Estimation of Macroeconomic Disequilibrium Models with Regime Classification Information. VII, 128 pages. 1987.

Vol. 289: B.R. Meijboom, Planning in Decentralized Firms. X, 168 pages. 1987.

Vol. 290: D.A. Carlson, A. Haurie, Infinite Horizon Optimal Control. XI, 254 pages. 1987.

Vol. 291: N. Takahashi, Design of Adaptive Organizations. VI, 140 pages. 1987.

Vol. 292: I. Tchijov, L. Tomaszewicz (Eds.), Input-Output Modeling. Proceedings, 1985. VI, 195 pages. 1987.

Vol. 293: D. Batten, J. Casti, B. Johansson (Eds.), Economic Evolution and Structural Adjustment. Proceedings, 1985. VI, 382 pages. 1987.

Vol. 294: J. Jahn, W. Krabs (Eds.), Recent Advances and Historical Development of Vector Optimization. VII, 405 pages. 1987.

Vol. 295: H. Meister, The Purification Problem for Constrained Games with Incomplete Information. X, 127 pages. 1987.

Vol. 296: A. Börsch-Supan, Econometric Analysis of Discrete Choice. VIII, 211 pages. 1987.

Vol. 297: V. Fedorov, H. Läuter (Eds.), Model-Oriented Data Analysis. Proceedings, 1987. VI, 239 pages. 1988.

Vol. 298: S.H. Chew, Q. Zheng, Integral Global Optimization. VII, 179 pages. 1988.

Vol. 299: K. Marti, Descent Directions and Efficient Solutions in Discretely Distributed Stochastic Programs. XIV, 178 pages. 1988.

Vol. 300: U. Derigs, Programming in Networks and Graphs. XI, 315 pages. 1988.

Vol. 301: J. Kacprzyk, M. Roubens (Eds.), Non-Conventional Preference Relations in Decision Making. VII, 155 pages. 1988.

Vol. 302: H.A. Eiselt, G. Pederzoli (Eds.), Advances in Optimization and Control. Proceedings, 1986. VIII, 372 pages. 1988.

Vol. 303: F.X. Diebold, Empirical Modeling of Exchange Rate Dynamics. VII, 143 pages. 1988.

Vol. 304: A. Kurzhanski, K. Neumann, D. Pallaschke (Eds.), Optimization, Parallel Processing and Applications. Proceedings, 1987. VI, 292 pages. 1988.

Vol. 305: G.-J.C.Th. van Schijndel, Dynamic Firm and Investor Behaviour under Progressive Personal Taxation. X, 215 pages. 1988.

Vol. 306: Ch. Klein, A Static Microeconomic Model of Pure Competition. VIII, 139 pages. 1988.